PURPOSEFUL DIGITAL TRANSFORMATION

A PLAYBOOK FOR MODERNIZING YOUR BUSINESS

SURESH GOPALAKRISHNAN

Purposeful Digital Transformation

ISBN: 978-1-7353645-0-6

Printed in the United States of America

To Maya, Arjun, and Ashwin

who inspire me every day

Table of Contents

Preface i

PART 1 – The Why 1

Journey To The Digital Age 3

Customer Journey In The Digital Age 19

Set Your Digital Age Strategy 31

Handle Your Business Model With Care 49

PART 2 – The What and Where 55

Change Your Value Delivery Engine 57

Digital Technologies 67

Digital Infrastructure 77

Digital Operations – Art Of The Possible 97

PART 3 – The When, Who, and How 111

Map Your Digital Journey 113

People Strategy In The Digital Age 127

Lead The Change 135

Preface

Googling "Digital Transformation" returns more than 44,000,000 results. There is no quality or fact check on this information, and every article you read will offer a different definition for digital transformation. How are you going to make sense of this deluge of data?

The term digital transformation was coined with very good intentions. However, as with any trending buzz phrase, corporate marketing machines have jumped on and distorted digital transformation to market products they wanted to sell. This frenzy got out of control once analysts, consultants, and academics got into the fray.

So, what exactly is digital transformation? *The essence of digital transformation is changing the old ways of doing business to stay relevant and profitable in the digital age.*

The survival of your company in the digital age depends on implementing the changes required to stay relevant and profitable in the future. Yet, if you get swept up by the hype around digital transformation, you will waste time and money and worse – become irrelevant to your customers.

Welcome To The Digital Age

Over the last few decades, technology advances have ushered us into the digital age. The *digital age* signifies that we now have multiple generations who have grown up with copious amounts of computing, connectivity, and information at their fingertips. Many new technologies are shaping how we buy and interact with consumer products and services. As a result, your employees and customers expect consumer-like experiences when they interact with your company, your products, and your services even if you do not sell directly to consumers.

Even small businesses will have to adapt to changing customer expectations. Gone are the days when you could run a hole-in-the-wall restaurant that thrives on word-of-mouth. You better have online ordering, partnerships with popular delivery services, and exceptional customer reviews (the new word-of-mouth). Of course, if the food is terrible, digital transformation is not going to help.

In the digital age, it is not technology that is forcing businesses to change. It is customer expectations that are forcing changes in how business is done. This significant piece of wisdom is obfuscated by the hype surrounding digital transformation.

As you plan your digital transformation, you will find that you increased the value to the customer but decreased the value to your business. Shifting value from the business to the customer is not unique to your business. Most transformations to digital business models end up with a lower profit compared to the starting point. The lower profit expectation is what allows those who attack your business to gain market share against you.

Attackers with new business models transfer value to customers hoping that their product gets a significant share of the market. These attackers are not expecting to generate the same profit as you, and have no business to lose. Having even one attacker with nothing to lose is a dangerous situation. If you do not take care of your profit, your digital transformation will be a disaster.

To prevent erosion of value to your business, you must think of digital transformation as *updating your business model to increase value to customers, reduce cost, AND create new sources of revenue.*

Purpose Of This Book

It is challenging to make consistent profits in the digital age because of the constant disruption of profit pools. At the same time, digital transformation will mean different things for each of your functional experts.

Your CFO may propose to automate processes and offshore them. Your CMO may want to buy more online ads and digital marketing tools. Your Sales VP may want to use a new CRM tool. Your CIO may propose to upgrade the entire IT infrastructure. All of these may have a place in your digital transformation. However, the question to answer is: How do you select and execute the minimum number of changes for relevancy and profitability in the digital age? This book is a playbook that will help you purposefully transform your business for the digital age.

What Is In It For You?

This book will help you answer the question: How do I make the required changes to stay relevant and profitable when there are so many moving parts in the business?

There is no guarantee that everything I write in this book will be directly applicable to you because digital transformation is situational. It also not easy to read examples of digital transformation and translate them to your situation. You must tune the approach you take to your realities by getting back to basics:

- **Why** do you have to change?
- **What** should you change?
- **Where** should you make the changes – in your business or outside?
- **When** should the changes be scheduled?
- **Who** has to drive the transformation?
- **How** do you drive the transformation?

Starting with these fundamental questions will equip you to frame your situation and lead the change. You will be able to see past the hype, understand the impact of digital trends, and map your digital journey.

How To Use This Book

There are three parts to this book.

Part 1 – Focuses on the *Why*

Part 2 – Focuses on the *What* and *Where*

Part 3 – Focuses on the *When, Who,* and *How*

All company, product, and service names used in this book are for identification purposes only and are the property of their respective owners

PART I

The Why

CHAPTER 1

Journey To The Digital Age

We are carrying the customer experience we get in our personal life as consumers into the B2B (business to business) world of our work. Our expectations span the whole customer journey, from how we buy to how we dispose of a product.

Today, many generations have grown up with easy access to computing, connectivity, and advanced technologies. These technologies are potent enablers for your competitors and potential industry disrupters. These attackers will wield digital technologies to steal your customers and plunder your profits. They will provide more value to customers by offering:

1. Products that are better than yours
2. Lower cost of ownership for their products
3. Consumer-like experiences in the B2B world

To stay relevant and profitable today, you must offer more value to your customers than others. It is relatively easy to deal with incremental product and cost improvements provided by a competitor, but a significantly better or cheaper competitive product will force you to change how you operate your business.

It is important to note that in the digital age, it is value to the customer, not technology, that will force companies to change how they operate.

Digital Age Customer Expectations

Product match and total cost of ownership are essential components that customers value. However, to thrive in the digital age, B2B companies must also understand the extras that digital-age customers value. Customers have the following expectations[1] when they do business with a company. Companies that meet these expectations will be more relevant to customers.

Mobile

We do everything while we are on the move. We whip our cell phones out to check stock prices, order online, track a package, pay for things, watch a how-to video, and pay a friend for dinner. A large percentage of people consume most of the information they need on mobile devices. Mobile interactions are the most common way for consumers to deal with businesses. And this trend continues to grow stronger.

Everything as a Service

For a while, ownership was the only way to get access to some things, so we went along with the only option and bought items to get access to them. In the digital age, we own only the things that we have an attachment to or do not make financial sense to access as a service. We buy music, software, and even transportation as a service.

Search

We start our buying process with a search for products that can help us do what we need to do. We find almost everything we buy by searching on Google, Amazon, Walmart, and other marketplaces. These search engines not only find us the products we search for but also give us recommendations based on our searches. If a product or company does not show up in our search result (with or without the company paying for it), it might as well not exist.

Personalized

We want everything personalized. We don't buy bundles of products; we want to buy precisely the parts of the bundle we want. We want to personalize drinks from vending machines, cars with our choice of materials, and our clothes to fit us without going to a tailor.

Secure

We want companies to protect the data they collect about us. We want every one of our financial transactions to be secure and not hackable. We don't want companies to collect information from our devices without telling us. We also want the connected products we buy to be secure.

Fast

We want instant gratification. One day delivery is not enough. We want everything right away. How about two-hour delivery? We want to visualize how a paint color will look on our bedroom wall. We want to see how a piece of clothing will look on us without going to a store.

Free

We want things to be free to use. We will trade our privacy for free stuff. We will ignore ads to access free services.

Cheap

If we cannot get something for free, we want it to be as cheap as possible. We want to pay only the least amount that others paid for the product or service. We buy only after comparing prices. We search for and apply online coupons at checkout.

Social

We check online reviews before purchasing everything. We look at items that others purchased with the items we buy. We check Pinterest and Instagram for how people use products we are considering. Social interaction is so prevalent that we get suspicious if we cannot interact with a company. For example, we are skeptical of products with no reviews.

Transparency

We want companies to be transparent about their environmental impact and supply chain. We will choose environmentally friendly companies over those that are not. We are getting more conscious of how companies operate manufacturing plants and supply chains. We do not want companies to use child labor, bonded labor, or poor working conditions.

We support companies that contribute to social causes. For example, we want companies to prove that they and their suppliers do not use conflict minerals or unfair trade practices in manufacturing products. And a lot of retail and online companies now allow us to make micro-donations by rounding up our purchases.

We also want companies to be able to trace our purchases from the source to our doorstep. For example, consumers have started preferring food suppliers and restaurants who can trace their products from the farm to the table.

B2B customers have more advanced transparency requirements than those of consumers like us since they must meet their customers' transparency and traceability expectations.

A Real-life Example Of Digital Attacks

The music industry is a great example of how disrupters enabled by technology provided incremental value to the listener and changed an industry.

From concerts to vinyl

Once upon a time, the only way to enjoy music was to go to a live concert. Then music enjoyment got a lot more convenient when we could buy vinyl records. We could listen to music we owned when we wanted. Then record labels moved from selling a popular song on a "single" record to many songs on an "album" record. Although we did not like all the songs on the album, we bought the album at a higher price than a "single" because it was our only option. Record labels needed a way to distribute music, so music stores started popping up everywhere to sell music.

The original streaming service

Record labels next partnered with the original "streaming" service – the radio, to grow revenue. Record labels were paid every time the radio station played a song, and the radio stations, in return, got music as content to attract listeners. The more listeners a radio station had, the more the station could charge to air ads. This increase in the ability to charge more is due to the network effect – more participants in a network, the more valuable the network becomes.

Music lovers who could not afford (or care) to buy records listened to the ad-supported radio service. At first, these customers did not have a choice of the songs they got to hear from a station. But it was free, and they had to listen to the ads. Then some stations offered a call-in service for listeners to request songs they wanted to hear. When radio stations started the request line, it opened up a new symbiotic relationship between the music industry and the station. Record labels, for a fee, could "request" the radio station to play a particular song many times a day. Essentially, advertise the song more than other songs. Record labels could then influence the popularity of certain songs and artists to sell more of those albums.

From vinyl to CDs

While the artists and record labels enjoyed profits from vinyl records, the listeners were dealing with many annoyances. They had to carry records and record players around if they wanted to hear their music somewhere other than their home. Customers also started hearing the loss of music quality due to records wearing out.

When improvements in magnetic recording allowed tapes to be made in the form of compact cassettes, record labels started using cassette tapes to distribute music because they were cheaper and easier to make. Music listeners got the convenience of a compact medium that was playable using small portable players.

Record labels started to see music pirates when cassettes were used to distribute music. Pirates were able to copy music from vinyl records or cassettes on to a cassette. The average listener did not go through the trouble of copying music to cassettes, and record labels were able to shut down large scale music piracy.

The quality of music from cassettes also decayed with use, so record labels started using digital methods to record music on CDs. Pirates could play music from CDs and copy on to cassettes or CDs. But piracy was still only a small issue because the distribution of pirated music was difficult.

Digital attack – Round 1

Then came the digital attack. A series of non-related digital technologies – computers, email, digital music playing apps, and electronic bulletin boards – gained popularity among listeners. Digital listeners could now easily copy a song and share it with friends using email or bulletin boards. Soon record labels were taking legal action against digital pirates – individuals, bulletin boards, email services, Napster, etc. Large scale legal actions were not going to be a sustainable strategy for the record labels. But online music stores like iTunes and others offered them a way out, rights-managed digital music distribution.

A digital solution

iTunes offered a lot of value to listeners. They could buy and download an individual song for less than a dollar on the iTunes store. The song was cheap enough for most people, and a song buyer could play it on a few devices, so people did not care to go through the hassle of copying songs. Listeners could also send a link to the song on iTunes if they wanted others to know about it.

Listeners did not have to buy the album bundle, and album sales went down significantly. However, record labels and artists got more customers buying songs. As music sales moved online, most music stores went out of business.

Digital attack – Round 2

Radio stations were still not disrupted. That happened when a new set of music streaming companies came into the picture. With the availability of cheap bandwidth, the music streamers allowed listeners with internet access to listen to music anywhere for free, create and share playlists, and listen to playlists curated by experts. The record labels promptly sued the streaming companies.

Eventually, the streaming companies and record labels settled their differences with licensing deals. By the time the record labels and the streaming companies came to terms, free services became extremely popular. At this point, the network effect came into play, and streamers became attractive to advertisers. Now the radio stations had serious competition.

Streaming services, which allow listeners to personalize their music streams, have started to take more share of the ad dollars from radio stations. And radio stations have started to stream their programs using the internet to keep customers. Music streaming services now offer ad-free streaming for a monthly fee. Subscription streaming reduces the dependency of streaming services to ads. Customers enjoy subscription streaming since they don't waste time on ads. However, subscription streaming is a difficult concept for radio stations to implement due to public utility regulations.

Here come the digital giants

As the battle for ad revenue rages between the streamers and radio stations, these competitors are now facing another threat. The record labels found a new revenue source, a few large online-networks that can stream music. Enter Amazon, Walmart, and others.

These large online-networks have a large number of users, and the internet-based streaming users are part of one or both of these networks. These large networks are a threat to all companies that stream music on the internet. Walmart and Amazon will offer cheaper services because that is in their business DNA. And this will lower subscription fees for the whole industry. For example, Amazon bundles ad-free music streaming at no extra cost to Amazon Prime subscribers.

The Cast Of Characters

The evolution of the music industry was a long story with many plot lines. Understanding how the changes impacted the characters will give you clues to lead a purposeful digital transformation. Let us unpack the story from the view of the characters.

Artists

Early in the story, artists made money touring the world, offering concerts. Then their revenue shifted to album sales from record labels and, to a smaller extent, shows. Artists have now turned to concerts to recover revenue lost due to digital music distribution. Digital distribution freed artists to distribute music without record labels as intermediaries. It is now common for well-established artists to set up private record labels to promote and distribute their music.

Record labels

Initially, record labels made revenue by selling albums through music stores. Record labels then added a new revenue source from the radio stations. Life is good for record labels until they run into digital piracy. Rights-managed digital music distribution came to their rescue as they battle digital piracy. But this rescue came at the cost of lower revenue due to album unbundling and lower margin due to the low price per song. Music streamers became a new revenue source, but the streamers will cut down radio revenue. At the end of the story, large online-networks offer a new streaming revenue source but at a lower margin. The margins will be lower because these large networks know their value and are hard negotiators. The large networks may kill off some of the streamers and put pressure on record label profits.

Music stores

Well, they did not survive. Rights-managed music distribution solved the music piracy problem, and customers started buying music online. The instant gratification of online music purchases and the ability to purchase single songs meant the music stores had no value to the customer. The intermediary went out of business as collateral damage.

Radio stations

The original streaming service enjoyed advertisement revenues until the internet-based streamers attacked. Radio stations most likely did not think of free music streamers as a threat to them until it was too late to react. Now they are in a battle with the streamers and the large online networks. The ongoing struggle with streamers and large online networks may force a lot of radio stations to be internet-only operations.

Internet-based music streamers

Companies like Spotify and Pandora gave the ability to personalize and play songs from an almost infinite song collection if there is an internet connection. Due to the near-ideal listening experience, streamers are taking ad dollars away from radio and may force some stations to be internet-based operations. However, Amazon and Walmart threaten the future of these streamers.

Music Listeners

Early in the story, we could not listen to music unless we showed up at a concert. However, what we really want to do is listen to any song we want, at any time we want, for free. We also don't want to own a limited set of music; carry our music around (records, cassettes, CDs, music files). Music streaming is now getting us close to our ideal listening expectations.

A Few Observations

As the music industry evolved, record labels thought of people who bought music as just that – customers who buy music. However, people bought music because it was their only option to listen to the songs they liked. But their real need was not to buy/own music but to listen to songs they want, anywhere, and anytime.

Music listeners did not think that they could listen to music without paying for it, so they did not ask for free listening. When radio came along, listeners could hear music without paying but could only listen to what the radio station played and had to "pay" by listening to ads. The music streaming services get listeners closer to their ideal listening expectations.

The record labels were also not proactively trying to meet listener preferences. They knew listeners did not want to buy albums. But changing that meant lower revenue, so they continued with the bundling until song piracy became easy.

One player in the music industry did not survive, the music store. When the music industry and customer preferences evolved, music stores had nothing of value to offer to customers.

These observations suggest it is good to do some introspection before changing how you do business.

- Do you know the "real" requirements your customers have for the job they want to do with your product?
- How do you find out the "real" needs if your customers are not able to articulate their needs?
- What makes you relevant to your customers now?
- What changes in the industry will make you irrelevant?
- Do you offer enough value to customers in the changed (transformed) industry?
- Are you missing opportunities to attack?

Endnote

1. Customers express the experiences they want when they deal with a company as their expectations and a business that provides the experiences they expect is of high value to customers. In the rest of the book, I will use customer expectation, customer value, and customer experience as interchangeable terms.

CHAPTER 2

Customer Journey In The Digital Age

Customers will have many expectations each time they interact with you. Therefore, a map of the customer's journey when they consume your product is essential to understand their interactions and expectations.

The Customer Journey

There are many ways to represent a customer's journey. For the purpose of this book, let's use a simplified version of the "consumption chain" described by MacMillan and McGrath[1]. A typical consumption chain maps the stages a customer goes through when they interact with you and consume your product. The journey starts when a potential customer becomes aware that they have a need that your product can solve and ends with them disposing of your product at the end of their need. The picture below shows the customer journey at a high level.

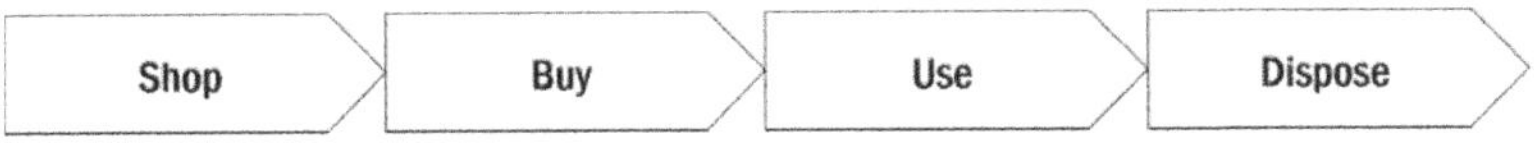

Shop

In this stage, the customer becomes aware they have a need (a job-to-be-done). Customers then proceed to search for something that will solve their need. After identifying many options, customers develop criteria to select the option that will work best for them. With the selection criteria in hand, potential customers will proceed to assess products and make their selection.

Buy

Most consumers buy things from online stores where they find the lowest price and free delivery.

Use

In this stage, customers start using your product and will have a new set of needs. They may need an expert from your company to help them with problems they face during use, store or transport the product, and maintain, repair, or return the product.

Dispose

In this stage, customers reach the end of their use of your product in its current form. They may upgrade, resell, repurpose, replace, recycle, or discard the product.

The following picture captures a map of the customer journey with the next level of detail. Customers value and evaluate all these interactions.

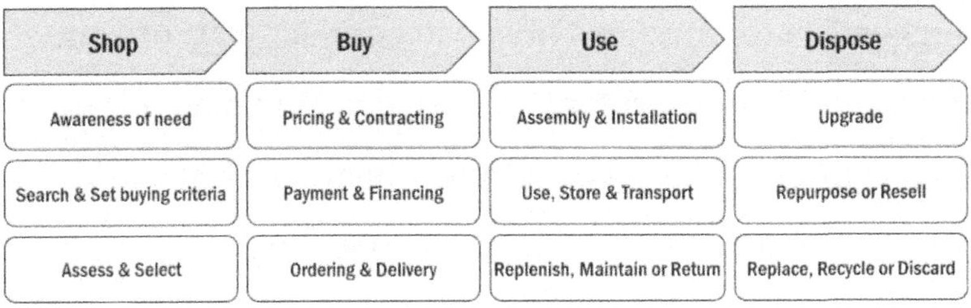

Shopping In The Digital Age

Awareness of need

The customer being aware of a need is the first step in the shopping stage of the journey. Some event usually triggers the awareness of a need. For example:

- Replacement – something broke down, and you need it replaced
- Life Event – you are going to have a baby, and need a car seat
- Advertisement – you saw an ad and realized you could use the product

In the digital age, all these examples are still valid, but there are more ways for us to become aware of needs. One of the most common digital awareness generators is recommendation engines (see Chapter 6). These engines suggest items that other people bought along with the product we are evaluating or buying. For example, when you are about to buy a washing machine, you may get a suggestion to purchase hoses to connect water to the machine. The particular washing machine may not have water hoses in the package, and you are now aware that you need water hoses to use this machine.

If you don't capture the customer at the moment of awareness about the need, they will start searching for a solution. And a search will display many competing products, which opens up the purchase to competitors. Therefore, capturing the customer at the moment of awareness is very important to your success.

It is imperative to understand the trigger events that make potential customers aware of a need that your product can solve. If you know the trigger event, you can start interacting with potential customers and get them to buy your product. Awareness of need is not the case for new customers only. An existing customer may need replenishing supplies when they use your product.

Consider the case of a personal printer reporting that ink is about to run out. If you allow the printer to send the "ink about to run out" trigger to its manufacturer, they now have many options to act on that trigger. For example, the manufacturer could incentivize you to buy ink or sign you up for a service to ship ink when they get the trigger.

In the digital age, companies collect and predict awareness triggers to capture customers at the moment of awareness or even before they are aware of the need. If you buy coffee from an e-commerce site, you will notice some examples of this. The supplier will typically offer the following:

- A 5% product discount if you subscribe to get coffee delivered to you at regular intervals. The supplier knows that people who buy coffee buy it again, and they don't want you to do that elsewhere.
- If you sign up for a subscription, the supplier will suggest a subscription quantity and delivery interval based on the buying patterns of other customers. Of course, you can adjust these if you need to.
- If you don't sign up for a subscription, the supplier will monitor your purchases and come back to you with a new subscription offer based on the quantity and delivery intervals derived from your purchases.

Another approach is to predict needs before the customer is aware of the need. Even if a supplier does not offer a subscription service, they may offer reminder services. Based on their customer's re-ordering patterns, suppliers will remind you that it is time to replenish supplies. I am sure you get emails saying you bought some products six months ago, and you may want to re-order or stock up. It is also common for service companies to remind you that you may need service and send a discount coupon.

Search and set selection criteria

Potential customers may or may not have selection criteria when they start searching for a solution. Since creating the selection criteria is the next step, e-commerce marketplaces make a couple of comparison options available. In the first option, the customer can select a few items, and the site will create a comparison chart based on the important product specifications. In the second option, the site will put together a comparison chart of popular substitutable items, even if the customer does not ask for the comparison. You will see such non-requested, auto-generated comparisons on Amazon. The comparison chart in the second option is an attempt to offer options that the potential customer has not considered.

There are two caveats when using comparison charts provided by online marketplaces.

1. Since the online marketplace establishes the criteria, the chart may not have all the features you should compare across options to make a good selection.
2. Since online marketplaces allow suppliers to pay a fee and sponsor products, the comparable products displayed may not be the best or the most popular product. Instead, they may be products that suppliers paid the site to show in the comparison.

Assess and select

We don't rely just on auto-generated comparisons. Peer reviews have the most influence on our selection. We want to know how many "stars" customers gave to the product when they rated it. We diligently read customer comments that accompany ratings – especially comments of the reviewers who gave a single star.

Depending on the item they are buying, a customer may narrow down the options and decide to test their preliminary selections. Most digital-age businesses offer some sort of try and buy scheme. Examples include online trial, placing the image of an eyeglass frame on a customer's photo, or a 30-day in-home trial.

Buying In The Digital Age

Pricing and contracting

Digital-age customers get a variety of tiered pricing schemes. The one that you must worry about is the free product. Free products are not truly free. Customers pay for these products by being targets for ads. If your product has no compelling value over the "free" product, you will become collateral damage to the advertising company.

In Business to Consumer (B2C) transactions, consumers have no choice but to agree to the terms and conditions set by the e-commerce site. B2B transactions can be more complicated than B2C transactions because once B2B customers select an option, they will negotiate pricing and their contract before placing the order. But B2B buyers are B2C buyers in their personal life, and this overlap between purchases for home and work means they value all the conveniences they have as B2C buyers.

Payment and financing

Digital-age consumers get a variety of payment and financing options at the time of purchase. Some common options are installment payments, finance a loan at the point of sale or checkout page, pay-per-use, and monthly or yearly subscriptions. And you can also pay using credit cards, direct transfer, and PayPal.

Ordering and delivery

Most consumer orders are placed online. The next step after ordering is product delivery. As consumers, we get instant downloads and same-day delivery. We can also get stuff delivered on a specific date. B2B customers expect these conveniences as well.

Product Use In The Digital Age

Assembly and installation

In most cases, consumers just have to open the package, and the product is ready for use. These days we get annoyed if the product requires charging for a few hours before use. We get even more annoyed if we get a product and then realize we must buy a battery before use. Some consumer products, especially furniture, need assembly and installation on-site before use. Consumers have the option to buy assembly, installation, and services with varying levels of service at tiered pricing.

Use, store, and transport

Most consumer products are intuitive to use. If we need to figure out how to use a product, we go to the product website or YouTube to watch a how-to video. When we start using the product, our needs shift to solving issues we run into during use. We use self-help options in case we run into any problems while using the product. These include knowledge bases, videos, and user forums. If we cannot solve the problem ourselves, we have the options of chat, email, or phone support.

Services that store and transport certain types of products can also be of great value to customers. Heavy or dangerous items that customers have to move among their sites are examples of products in this category. And occasional-use equipment that occupies a lot of space are candidates for storage or subscription services.

Replenish, maintain, and return

During the product use stage, many service-related needs pop up. The product may need replenishment of supplies, regular service for proper operation. As consumers, we get reminders for service and the ability to book service appointments online. Our suppliers make it easy to send products for repair and even to return products that do not meet our expectations.

Product Disposal In The Digital Age

As consumers, we want to maximize the value of the product we purchased. So, we search for ways to repurpose products. If we fail to repurpose a product, we try to sell it in the used market. If we find no buyers for the used product, we try to feel good about disposing of it by donating it. If we cannot give it away, the next feel-good disposal is to recycle the product.

A significant development in the last few years is the acute awareness of the impact of product waste on the environment. This awareness is leading us to think about the ease of product disposal during the selection stage. And our expectation has evolved to not dumping in the landfill. Therefore, companies that help up dispose of products we stop using are of value to us. For example, printer companies routinely allow us to ship used cartridges back to recycling centers.

The Road Ahead

Since it is best to capture the customer at the moment of awareness, digital-age businesses monitor customers and potential customers by collecting and predicting awareness triggers. If they can't capture the customer at the moment of awareness, they make it easy to find their products and comparison shop when customers search for solutions. Winners influence the selection criteria allowing comparisons on their site with benchmarks they select and know will favor their product.

Friction-free buying, free shipping, community-based support, hassle-free returns have all started showing up in the B2B customer journey. On the B2B side, companies have started specifying environmental requirements for making, shipping, and disposing of the product. Digital-age B2B buyers like these conveniences and expect suppliers to be environmentally friendly and be good corporate citizens.

Endnote

1. MacMillan, Ian & Gunther McGrath, Rita. (1997). *Discovering New Points of Differentiation.* Harvard Business Review. 75. 133-8, 143.

CHAPTER 3

Set Your Digital Age Strategy

If your business has been around for a long time, you probably established the business strategy a long time ago. Business strategies decay every day, and what was once a great business strategy may not be so great now. In the digital age, this rate of decay has accelerated due to:

- Changing customer expectations
- Technological advances
- Investment in new entrants who disrupt industries
- Huge, tech-savvy companies who can get into your business with ease

Given the pace of change in the business world, if you don't update your business strategy for the digital age, your business will fail because it will become irrelevant to customers.

You can make many adjustments to defend your position, but that is never enough. You never win a game by only playing defense. A mix of offense and defense is necessary to win the game. The same is true with your business. You have to use a mix of defensive and offensive changes to your business strategy to stay relevant and profitable in the digital age. These changes are the *Why* of your digital transformation.

While it is very easy to say, "make changes to your business strategy," it is not an easy task, especially if you are not clear about the difference between your business strategy and your business model.

Business Strategy vs. Business Model

A "business strategy" is the selection of customers to serve, the value to offer, countries of operation, etc., whereas a "business model" is a combination of how a business delivers a product of value to its customer and how the business makes money delivering the product. You first set your business strategy, and then execute it using a business model. Below are two very simplified examples to clarify the difference between the two.

Example 1: Walmart

When Walmart started, it sold to customers in rural areas. The business model delivered the following value to the customer – a wide range of products at low prices. In the next step of its revenue growth, Walmart decided to serve customers in cities. City customers are a new type of customer, so this is a change in business strategy. Since the value delivered to the new city customers was the same as the one for existing rural customers – a wide range of products at low prices – Walmart did not have to change the business model and could reuse a large number of business processes. The business model has since evolved since Walmart is very good at adopting new technologies.

Now consider the move Walmart took to defend against a digital attacker – Amazon. Walmart decided to serve online customers. Is this a change in business strategy or business model? Walmart added new online customers and provided new value to existing customers. This a change in business strategy. Walmart then changed its business model by adding new business processes like online ordering and shipping.

Example #2: LinkedIn

LinkedIn's initial business strategy was to serve two types of customers – professionals and recruiters. The value proposition to professionals was the ability to post resumes online for "free." The value offered to recruiters was that they could access a resume bank for a fee to find candidates. The business model was designed to operate web interfaces for both professionals and recruiters.

LinkedIn then added job advertisements to its paid products allowing recruiters to use LinkedIn to do another job – advertise their jobs to potential candidates. Allowing customers to do a new job is a business strategy change. Most business processes remained the same, but a new set of processes were added to allow placement of ads, find potential candidates, and display ads. Services like content sharing and training added by LinkedIn are also business strategy changes that incrementally added business processes to its business model.

Vulnerability Assessment – Knowing When To Defend

Much like Walmart in the first example, your business can be vulnerable to new entrants (Amazon) or existing competitors taking advantage of digital capabilities to offer more value to customers. The first step to identify if you have to defend is to do a vulnerability assessment. The goal of the vulnerability assessment is to identify the gaps between the value you deliver, the value your competition delivers, and the value that customers expect and which of those gaps will significantly hurt your business. Vulnerability assessment is complex because even with the simplified customer journey map in

Chapter 3, there are 12 steps (four stages and three steps per stage) that you must assess.

You can use the following steps to check if your business is vulnerable to attack:

Step 1: *Identify the customer expectations in each stage of your customer journey map.* Customers signal their expectations, so you just have to collect the signals using a systematic process. Observing your competition will also allow you to learn a lot about customer expectations. Use the following techniques to collect customer expectations.

- Collect customer complaints and requests filed with your support team – These requests and complaints usually point to gaps between your product's functionality and what the customer wants to do.
- Capture what customers are telling your sales teams – Are they asking you to unbundle your products? Are they asking for free shipping?
- Interview customers – Ask them what they are trying to do with your product? Walk them through each stage and step of the customer story and ask how you can make each step less painful. Ask them about the value and conveniences provided by competitors for each task within each stage of the customer journey.

Step 2: *Find out how well your competitors address the expectations.* Conduct competitive analysis and customer interviews to identify the value and convenience offered by competitors in each stage of the customer journey. Who does it well? Who does a mediocre job? Who does not meet expectations, and why?

Step 3: *Combine the results of Steps 1 and 2 into a comparison table.* This table will show the customer expectations and how well competitors address them.

Step 4: *Add entries to the comparison table to show how well you meet the customer expectations.*

Step 5: *Identify gaps that make you vulnerable.* Figuring out which gaps in each stage of the customer journey makes you vulnerable to attack is difficult because you will find many gaps between the value you deliver and:

1. The customer expectations
2. The value delivered by the competition

The Blue Ocean chart described by Kim and Mauborgne,[1] can be used to identify gaps that will make you vulnerable. This chart will come in handy to figure out your offensive moves as well. The next picture shows a sample comparison map between you and one of your competitors.

Your real-life chart will have more competitors and expectations.

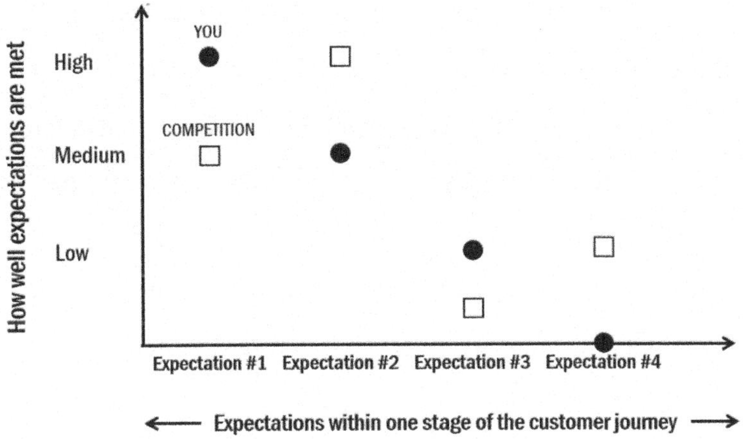

Once you generate a map, as shown above, you can assess which gaps make you vulnerable by asking the following questions:

1. What will happen to revenue, profit, and customer relevance if we don't meet Expectation #2 and #4 at the same level as the competition?
2. Why are we not doing anything about Expectation #4?
3. Why are we not meeting Expectation #2 at the same level as the competition?

The answer to the first question will tell you if the gap makes you vulnerable. The answer to the second and third questions will inform you of the changes you will have to make to provide this value to your customers.

During the discussion with your team, be sure to ask the following additional questions:

1. If meeting Expectation #2 and #4 lowers profit, can we reduce enough costs elsewhere to recover the profit loss?
2. What will happen if we drop our delivery on Expectations #1 and #3 to the same level as the competition?

The answers to the two questions above will give you ideas for cost reduction changes to recover profits.

Formulating Your Defensive Moves

Don't let the perfectionist in you drag you down the path of defending every gap you find. Since your team will be able to execute only a few big changes properly, you must pick the changes you make to give you significant benefits.

Start by picking value gaps that will lower revenue or relevance significantly and then use the loss of profit as the criteria to refine your defensive changes further.

Pick #1: *Changes with the least profit loss.* Anything that increases the value to the customer without reducing value to the business is worth doing and will increase your relevance. Address the gap and bring the value you deliver at least up to par with the competition. For example, your policy preventing no-hassle returns may assume that a lot of customers will abuse the convenience. However, if you believe that customers will not abuse the convenience, you can offer no-hassle returns without loss of profit.

Pick #2: *Changes with short-term profit loss that can be recovered in the long-term.* If you can deliver a highly valued expectation and recover the profit loss, offer the value immediately, and diligently recover the loss of profit. Time is of the essence.

Pick #3: *Changes that cause long-term, non-recoverable profit loss to address a value provided by all competitors.* Unless your product is significantly better than your competitors', you will eventually feel the pain of not delivering this value. For example, if all your competitors are offering two-day shipping, you will have to offer it.

Pick #4: *Changes that will significantly reduce your profits.* If your attackers are driving your industry profits lower, your choices are to accept the lower profit, differentiate to command a premium, change your business strategy (new customers, new jobs-to-be-done), or exit the business.

Formulating Your Offensive Moves

There are two steps in formulating your offensive moves.

Step 1: *Identify how well your industry meets customer expectations and figure out if you can change your business model to provide value to customers that is better than the industry norm.*

Step 2: *Identify new value and conveniences you can offer to customers that no one else offers. Use these new value and conveniences to attack.*

You can reuse the preparation you did for the vulnerability assessment as the starting point for your offensive moves.

Step 1 in Action

The basic question you have to answer in this step is: Will improving your business model's delivery on any of the customer expectations make your business more relevant to customers?

Let us walk through attack formulation using the customer expectation comparison chart that was prepared for the vulnerability assessment.

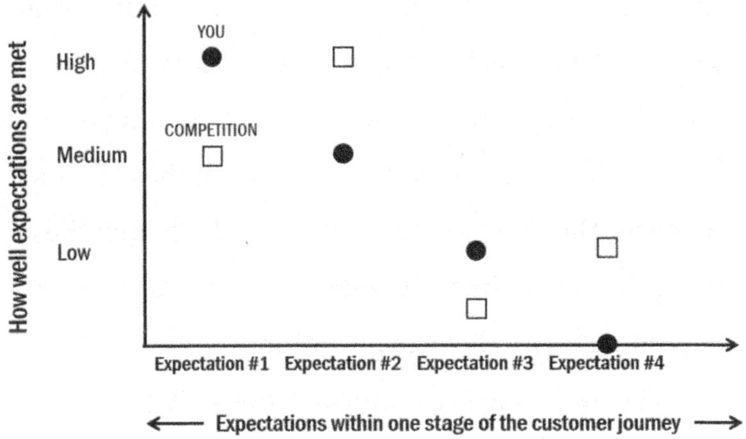

- For expectations like Expectation #1, where you deliver more than the competition, and at the highest level that the customers expect, the question to ask is: Can we enhance this value beyond the industry norm to make a difference? If the answer is yes, attack with these expectations.
- For expectations like Expectation #2 and #4, where you deliver below your competition, improving your performance to the same level as the competition is a defensive move. If you want to attack using these expectations, you must ask the question: Can we enhance this value beyond what the customer expects

and make a difference? If the answer is yes, attack with these expectations and outperform the competition.

- For expectations like Expectations #3, where you deliver more than the competition, but both you and your competition are still short of what the customer expects, the question to answer is: Will taking these experiences to the next level help? If the answer is yes, attack with these expectations.

The following is a pictorial view of Step 1.

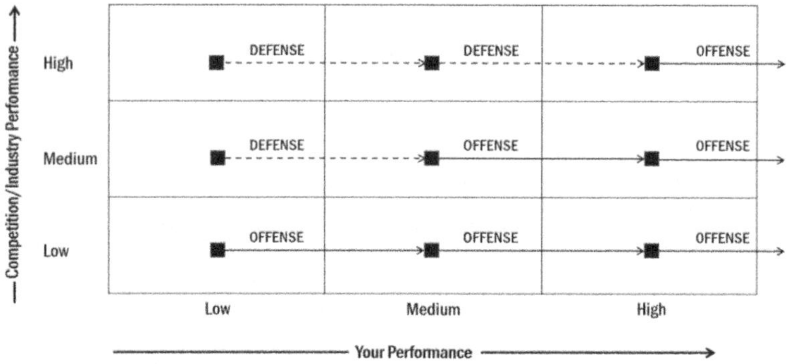

You will identify far more potential changes than you can execute well. As with defensive moves, you must prioritize the changes you make. You should pick the least number of offensive moves in each stage of the customer journey and turn them into significant attacks.

Start by picking value and expectation gaps that will increase revenue or relevance significantly and then use the loss of profit as the criteria to refine your offensive changes further.

Pick #1: *Changes with the least profit loss.* Expect that your competitors will follow suit immediately.

Pick #2: *Changes with short-term profit loss that will be extremely painful for competitors to copy.*

Pick #3: *Changes that will bring down the industry profits.* If you are the underdog and losing relevance fast, then take the industry profits down and offer all sorts of value to customers. If you choose this path, you must figure out ways to reduce operating and product costs to thrive with the lower profits.

Step 2 in Action

The main question to answer in Step 2 is – Can you deliver features or experiences that your industry is not providing but can be of great value to customers? Expectations #1 through #4 are of no help to you in this step. However, the insights you develop about customer expectations in your vulnerability assessment will come in handy in this step.

Do you see any opportunities to go beyond what the industry is doing? Can you pull away from the industry average business model and differentiate yourself? You may not need a new product feature or convenience to mount an attack; contribution to an environmental or a societal or cause may be enough.

The following is a pictorial view of Step 2.

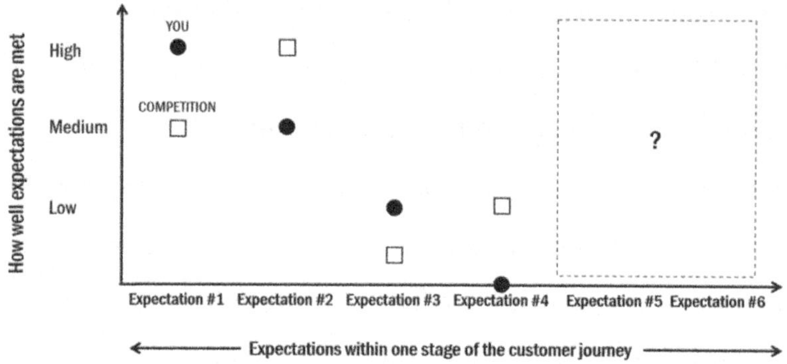

Remember to consider the impact of each change on your profit when you select changes to execute. You can apply the selection method used in Step 1 to sequence your attack. When you find something that you can attack with, do not wait for perfection. Go to market as soon as you have a minimum viable solution.

Accelerate Your Digital Strategy Setting

Changing the value offered to customers, the *Why* of your digital transformation drives business strategy changes. If you believe your business is under attack, then speed up your strategy setting using the consumer expectations list below. For your convenience, I have mapped the expectations to the stages of the customer journey and transparency needs. Customize your list by adding specific items that are of concern to you.

Shop

- Value is well articulated
- Easy access to product details
- Help in establishing selection criteria
- Ability to compare with competing products
- Customer testimonials
- Peer reviews
- Access to budgetary pricing
- Free shipping
- Don't want to buy a product bundle
- Product security

Buy

- Pricing options – per-use/transaction/outcome, free
- Payment options – subscription, per-use
- Payment types – contactless credit card, direct transfer, PayPal
- Financing options – lease, installment plan
- One-click ordering
- Financial transaction security
- Automated order checking for accuracy

- Recommendations for items that are missing from the order but are essential to use the product
- Delivery options – one day, two days, scheduled

Use

- Assembly available at the source or on-site
- Installation available on site
- DIY videos on how to assemble and install
- Usage model – own, as a Service, on/off-premise
- Problem handling – self-service, chat, email, phone
- Minimal disruption to repair and service the product
- Hassle-free product return

Dispose

- Ability to upgrade without disrupting use
- Easy replacement of product – same form with the same or better fit and function
- How-to instructions for repurposing products
- Facilitation of used product resale
- Product trade-in
- Facilitation of used product donation
- Recycling support
- Low/No environmental impact when disposing
- Hazardous material disposal support

Transparency

- Transparency on manufacturing, shipping, and disposal – conflict minerals, hazardous substance, environmental impact
- Transparency on labor practices – no child labor, no bonded labor, worker safety
- Trace products from source to the customer site

Endnote

1. Kim, W. Chan., and Mauborgne, Renée. (2005). *Blue Ocean Strategy: How to Create Uncontested Market Space and Make the Competition Irrelevant.* Harvard Business School Press. Print.

CHAPTER 4

Handle Your Business Model With Care

Digital transformation implies you are going to change your way of doing business to be relevant and profitable in the digital age. The changes you make to stay relevant will impact how much money your business generates. Therefore, mapping and understanding how you make money are essential to staying profitable when you embark on a digital transformation.

As mentioned in Chapter 3, the business model is a formal way of capturing how a business makes money. The components of your business model determine the *What* and *Where* of your digital transformation.

Revisiting The Business Model

Let's use a simplified version of the business model as defined by Johnson, Christensen, and Kagermann[1]. At a very high level, a business model has four components:

1. *Product:* The business delivers a product to customers for them to do a particular job. The business chooses the customer, job to be done, and the value to offer as part of the business strategy.

2. *Business processes:* Procedures used to design, market, manufacture, sell, deliver, and support the product.
3. *Resources:* People, money, policies, partnerships, etc. are used to execute business processes.
4. *Profit formula:* This formula is used to calculate how the business makes money. To simplify the discussion, let us assume that you have only one product. In this case: Operating Profit = (Unit Volume x Unit Price x Unit Margin) – Cost of doing business.

The following picture shows the relationships between the business model components.

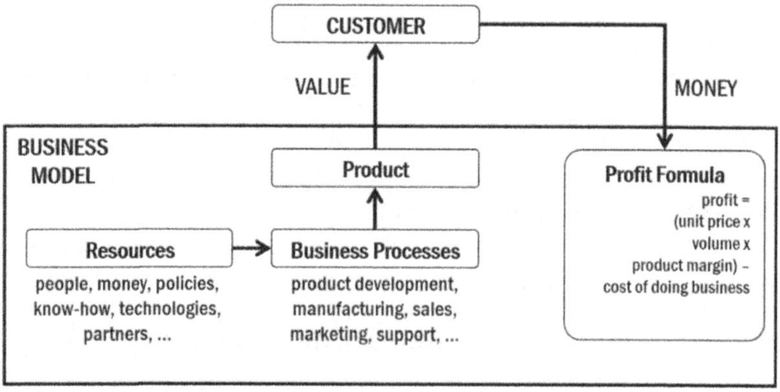

Your resources operate your business processes to deliver the product to your customers. Customers, in turn, value the product based on how well the product allows them to do the job they want to do. Customers pay you based on the value they perceive. You calculate your profit or value to the business using your profit formula. Your business model generates two kinds of value:

1. Value to the customer
2. Value to your business

The following picture shows how your business model delivers these two types of value.

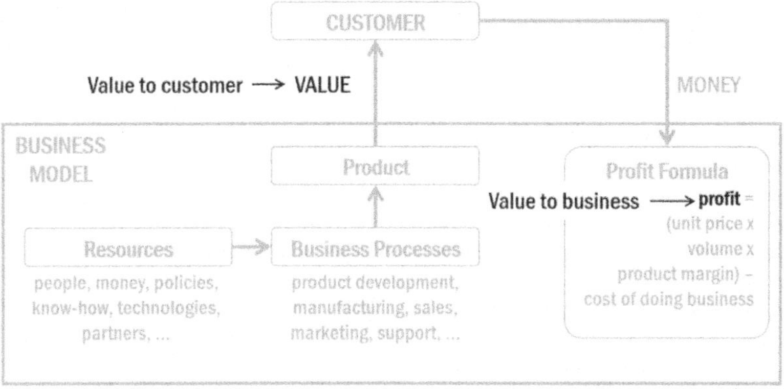

The price you command and the volume you sell and, therefore, the revenue you generate depends on how much your customers value the product. Customers highly value your product only when it allows them to do the job they want to do; in exactly the way they want to do it. When you consider the job customers want to do, think about the entire journey from how they shop to how they dispose of the product.

The following picture shows the relationship between the value to the customer and the revenue you can potentially generate.

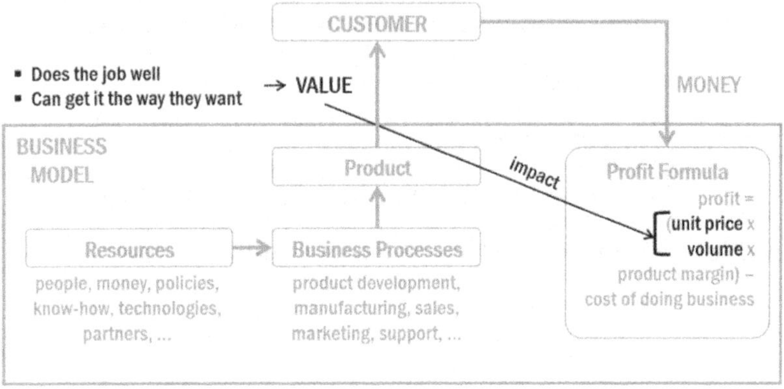

Your product cost depends on the resources and processes used to build the product. This cost, in turn, impacts the product margin. The cost of running the rest of the business processes is your cost of doing business. The following picture shows how business processes and resources impact the profit formula.

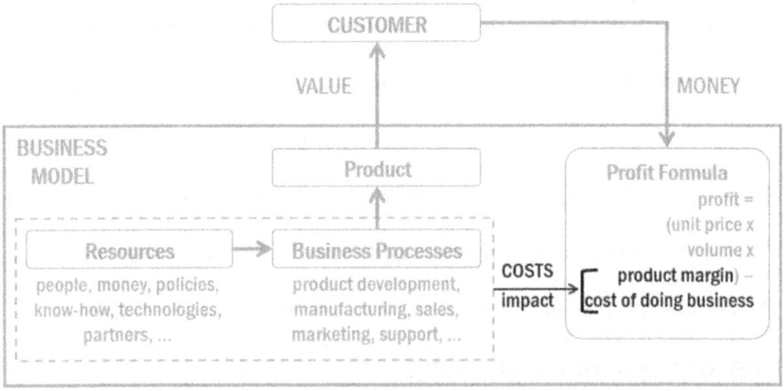

Cautionary Lesson

You will reduce the profit from your business if you make product and operational changes without deliberation. Here are a few examples of how profit can be impacted if you are not careful. Your profit will go down if:

- The side effect of a change reduces product capabilities or customer support
- The side effect of a change makes it difficult for customers to do business with you.
- You modernize your business processes. In certain cases, you have no choice but to do the process update. But it will cost you at least in the short term.

You can always attempt to price your product on a cost-plus approach to achieve a target product gross margin. This approach has two well-known issues:

1. Customers must perceive enough value from your product for you to command your target price. Otherwise, they will not buy it at the price you set.
2. If you do not know the value to the customer, you will not be able to capture all the value that should flow to your business.

The lesson to remember is – changes you make to the left side of the business model pictures impact the right side. A digital transformation is a balancing act of providing more value to the customer without destroying your business. So, handle your business model with care.

Endnote

1. Johnson, Mark & Christensen, C.C. & Kagermann, Henning. (2008). *Reinventing Your Business Model.* Harvard Business Review. 87. 52-60.

PART 2

The What and Where

CHAPTER 5

Change Your Value Delivery Engine

You make changes to your business model to implement the defensive and offensive changes for your digital transformation. The changes can be product features, removing impediments to shopping, pricing, etc.

So What Should Be Changed?

Customers interact with your product and your business processes and form opinions based on these interactions. These perceptions will determine your product's value in their mind, and if they think it is easy to do business with you. Since you deliver your product and the resulting value to customers using your business processes, you will have to change your business processes to meet digital-age customer expectations.

From Chapter 4, you will recall that your resources (people, policies, etc.) drive business processes. Therefore, when you make changes to your business processes, you may have to make changes to those resources as well. Let's look at how business processes are put together as the first step in identifying what to change.

The Business Process Stack

Business processes are built using the following layers.

- **Layer 1:** *Business Applications.* The software used to implement your business processes
- **Layer 2:** *Technologies.* A variety of technologies are the building blocks for business applications
- **Layer 3:** *Data*[1]. Business applications rely on data from inside and outside your company
- **Layer 4:** *Infrastructure.* The business applications run on your IT infrastructure
- **Layer 5:** *Business Policies*[1]. Your business policies govern how business processes are executed
- **Layer 6:** *Partnerships*[1]. Relationships with third parties that augment your business processes

The following picture shows how this structure maps into a stack diagram.

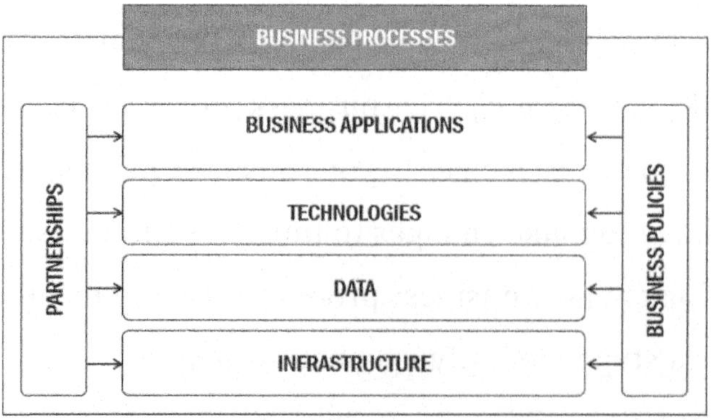

Now that we have the basic stack diagram for business processes, here are some examples of the components within each layer.

Layer 1: Business Applications

Business applications process data using business logic and take actions based on the outcome of processing data. Typical applications include:

- Customer Relationship Management (CRM)
- Enterprise Resource Management (ERP)
- Accounting applications
- Product development tools

The number of business applications that are used in a large enterprise will blow your mind. It is not unusual for a large enterprise to have thousands of business apps. An enterprise customer of mine had over 8,000 apps in use within their company.

You can either buy business applications from software vendors or develop the applications inhouse. If you decide to buy business applications, you may have a choice to buy a perpetual or subscription license and run the app on your computers or use a software-as-a-service

(SaaS) model in which the application runs on the SaaS providers' computers.

Requests from functional and business teams who own the business processes result in application development inside companies. These internal apps and apps customized by third-parties lead to the proliferation of business applications in a company.

Layer 2: Technologies

It is rare these days for programmers to develop an application from scratch. Programmers build apps using building block technologies. The use of these technologies speeds up the implementation of business applications. Some commonly used base technologies are:

- Databases (many types)
- Storage (many types)
- Web front end (webservers – for displays, forms)
- Load balancers (to handle a lot of requests)
- E-commerce engines (to support online purchases)
- Payment systems (interfaces to credit cards, banks)
- Programming languages
- Manufacturing technologies (silicon, packaging)

Layer 3: Data

The data you have accumulated in your business is very crucial for your business. Typical datasets in enterprises include:

- Customer profiles (preferences, past orders, payment details, support requests)
- Process details (status, efficiency, support requests)
- Supplier profiles (performance, quality)
- Infrastructure details (outages, costs)
- Inventory details (finished goods, work in progress)
- Marketing data (brand assets, databases, event calendars, leads)
- Sales (opportunity pipeline, leads, orders)

Layer 4: Infrastructure

Your infrastructure includes manufacturing facilities, computers, storage systems, communication networks, and logistics capabilities. However, for the purpose of this book, I will stick with your IT infrastructure. How you use computing and storage capacity is a topic for discussion in the future so let me define two terms:

- *On-premise (on-prem):* You own or lease equipment and use it in a facility owned/rented by you. You may have a server room or even several datacenters full of servers, storage, and networking equipment.

- *Off-premise (off-prem):* You rent computing and storage capacity from an infrastructure-as-a-service (IaaS) vendor – aka Cloud vendor. Although IaaS is a multi-billion-dollar industry, enterprise migration to the Cloud is in its infancy.

Layer 5: Business Policies

Your current set of policies (rules and metrics) are in place due to your team's accumulated wisdom and best practices from the industry. These policies set the rules of how the business operates. Payment and shipping terms, data access control, cloud usage policies, etc. are examples of business policies. But these policies can get in the way when you try to change business processes to meet new expectations and to provide more value to customers. Legacy policies will kill a digital transformation even if you implement all other changes flawlessly.

If you adopted new technology and infrastructure along the way, the changes to many of the layers are going to be incremental. And a lot of your business processes will already be digital-ready. If this is the case, you won't have to change business applications, technologies, or infrastructure to provide additional value to customers. You just need to change the policies associated with the business process.

Most of the time, policy changes can be done quickly and without spending money. Sometimes, you cannot quickly roll out a policy change because the change may cause revenue recognition changes. For example, if you change your policy from selling software licenses to a subscription-based model, you will change how revenue is recognized. Now the finance team has to educate your investors and warn them about the transition to avoid panic. Even in this example, you could move to the new policy with little implementation cost. But you will opt to not move fast due to the change in revenue recognition.

The ability to change policies inexpensively and quickly is the good news. The bad news is that when you change some policies, you can incur costs. For example, it will cost you money if you change your policy from charging for shipping to free shipping.

Layer 6: Partnerships

You don't have to do all the activities required to run business processes yourself. You can deliver great value to customers using partnerships. There are a variety of third-party products and services that you can use to manufacture, support, and ship your product. Until you have the appropriate scale for your business, you don't have to own or build capabilities that you can get from a partner.

So Once Again. What Must Be Changed And Where?

Business process changes fall into three categories:

1. Changes you drive to increase value to customers and to stay relevant to them
2. Changes you drive to reduce product or operating costs to stay profitable
3. Changes forced on you by unavoidable updates to technologies or infrastructure – these changes have a high opportunity cost

The following stack diagram shows the next level details of business processes. As you can see from the picture, you can change many components within a business process to execute your digital transformation.

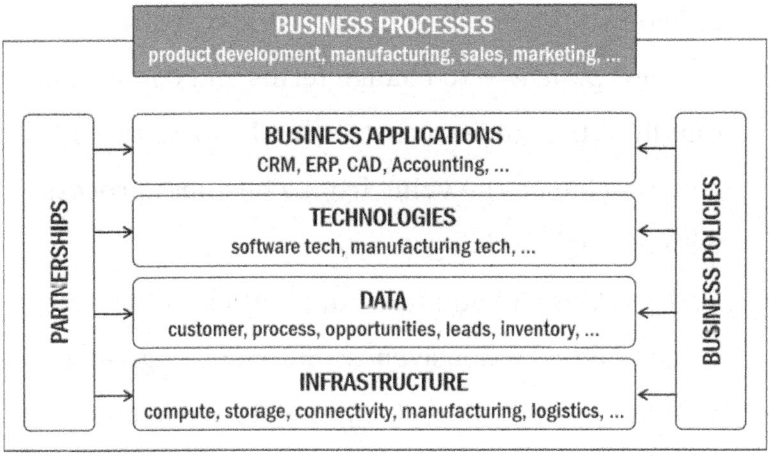

Now that you understand the business process stack, you can infer that business process changes drive changes to the following:

1. *Business applications* – to change a business process. You can develop these in-house or buy them from software vendors.
2. Technologies – to update or build new business applications.
3. *Data* – collect more of it to meet and exceed customer needs. Some of your data sources will be external to your company, so you may have to make changes, both internal and external to your company.
4. *Infrastructure* – to support business process changes and to add more capabilities to meet customer needs.

You can change your on-prem infrastructure or go outside your company to rent infrastructure.

5. *Business policies* – to change terms and conditions that limit the value to customers. In some cases, you can change the value from a business process with just policy changes.
6. *Partnerships* – to add new capabilities, get better cost, or better responsiveness. Since these agreements are with third parties, changes will have to be made outside your company.

Endnote

1. Data, business policies, and partnerships are very important topics. However, I did not dedicate separate chapters for them since the goal of this book is not to dive into every layer of a business process.

CHAPTER 6

Digital Technologies

An introduction to some of the critical technologies used in business applications of the digital age is essential to figure out how you can exploit them. Some technologies that we consider essential now have been in existence for decades. These stayed as niche research topics for a long time and became mainstream only after plenty of cheap computing came along in the 2010s. The building block technologies for the digital age are:

- B2B APIs
- AI Tasks – Machine Learning, Inference, Machine Vision, Natural Language Processing
- Robotics
- Big Data Analytics
- Payment Systems
- Near Field Communications
- Social Networks
- Search
- Blockchain

Basic technologies like web front-end, mobile apps, databases, and e-commerce engines are still important in the digital age. However, for the purpose of this book, I will cover technologies making their way from consumer interactions to the B2B world. Understanding these digital-age technologies will give you the options for a purposeful digital transformation.

B2B APIs

B2B Application Programming Interfaces (B2B APIs) allow businesses to exchange information between their computers. You can automate order placement, status monitoring, data querying, etc. using B2B APIs. Social media companies, supply chain companies, language translation services, and image recognition services all provide B2B APIs. Use available B2B APIs to automate your interactions with customers and partners. Internal processes, customers, and partners can automate tasks if your business applications (your own and 3rd party) support APIs. Automated tasks will, of course, increase agility and decrease cost for you, your customers, and your partners.

AI Tasks

AI tasks like machine learning, inference engines, machine vision, and natural language processing allow computers to figure out their surroundings and achieve predefined goals. These tasks were initially developed in the quest to implement artificial intelligence. However, these days a combination of these tasks is used to improve business performance.

Machine Learning

Machine learning is an AI task where computers develop a mathematical model from lots of data provided as input. Machine learning uses a variety of algorithms, including some that mimic functions of neurons found in our brains. Machines learn from past data, and there will be historical biases in the data. Biased data leads to biased models and bad results. Always remember that a "learned" model is only as good as the data used for learning. Ask your experts to investigate historical biases in your "learning" data and update learning with new data frequently.

Inference and Recommendation Engines

Inference engines use mathematical models developed using machine learning to identify conditions, make predictions, or make decisions. Inference engines have been around since the 1980s as part of expert systems. Expert systems have two components, a knowledge base, and an inference engine. The knowledgebase contains rules provided by human experts to handle specific situations and tasks. A simple inference engine could then detect the specific situation and apply the rules provided by the experts. An expert system is usually the basis for many chatbots. In new chatbots, the knowledge base is augmented with models generated using machine learning.

A recommendation engine is a specialized inference engine that makes recommendations based on learning past activities and trends. These engines are behind the recommended products displayed when you shop online.

Machine Vision

Machine vision is the AI task used to make a computer see what is in its environment. Optical and thermal sensors are the starting point for machine vision. Image processing algorithms clean up the image provided by the sensor. From these images, you can detect text, check dimensions, recognize people and things, etc., using image detection algorithms.

Business processes use machine vision paired with inference engines and expert systems. A common use case of machine vision is detecting defects during the manufacturing of products. Keep track of machine-vision-based apps as novel apps are becoming available at a rapid pace. Inference engines used with machine vision also use models generated by machine learning. Therefore, biased data can lead to bad assessments, especially in facial recognition.

Natural Language Processing

Natural language processing (NLP) is the AI task used to make a computer understand human language and respond using human language. Input to and outputs from NLP systems can be text or sound. A variety of algorithms that detect key concepts, sentiments, themes, topics, and contexts are the basis of NLP. The use of machine learning allowed NLP technologies to improve their accuracy in the last decade. You can detect trends (customer issues, Twitter feed, etc.) or create chatbots using an NLP system along with an inference engine.

Robotics

Robotics involves building machines that can execute tasks specified by a computer program. Robots can automate processes that are repetitive or threaten human health. Robotic process automation (RPA) lowers cost by reducing labor costs and improves quality by reducing human error. A lot of people incorrectly equate RPA to digital transformation. RPA and its enormous societal implications are off-topic for this book.

The word robot brings up images of human-looking androids and robotic car manufacturing lines. However, if you use the definition of "machine that can execute a task specified by a computer program," we can build robots without mechanical elements. So, RPA now includes software only robots that automate tasks linking many business applications. For example, a customer's email requesting service can automatically file a service ticket, order replacement parts, and invite the customer to schedule service. Software-only RPA provides better agility and responsiveness on top of cost reduction and higher accuracy.

Big Data Analytics (BDA)

Data analysis to derive insights to improve business performance, aka business analytics, has been around for decades. The hype around business analytics spawned business intelligence applications and the automated collection of data at every interaction. These days companies now collect data from every customer, employee, and machine to machine interaction. A large amount of data (Big Data) collected this way is tough to access and analyze. Analyzing big data needs special techniques, and these techniques are collectively called Big Data Analytics.

You will be a step ahead of your competitor if you can derive customer insights from your data. But in most companies, there is not enough data engineering talent to ensure the goodness of the data and data science talent to derive insights from the data.

Social Media

Social Networks like Facebook, LinkedIn, Twitter, and others. are becoming important for some of your business processes. Marketing, reputation management, customer support, and sales support are processes that can use social media feeds to improve performance. Companies use social media analysis to identify risks, sentiments, and trends. Social media analysis also drives business processes like stock trading and reputation management.

Search

Search is the primary research vehicle in the shopping stage of customer journeys. Search technology originated to search for information on the internet. However, now companies use search technologies to advertise and provide product information. Search technologies used inside companies make information access convenient for both customers and employees.

Payment Systems

Consumers are used to many payment options as online purchases increase, and retail stores adopt mobile payment technologies. New payment options include digital wallets, point of sale financing, automated e-invoices, direct bank payments, and prepay. Some of these payment options will become important for B2B companies as well. Most financial technology (fintech) companies offer these payment options for a fee, so you don't have to build them from scratch. Track fintech innovations to stay abreast of the latest in payment systems.

You can offer banking and financing to customers using business applications that integrate fintech. Besides new revenue streams, banking and financing increase customer stickiness and profits.

Near Field Communications

Near field communication (NFC) allows two devices to communicate over a short distance (inches) using radio waves. NFC offers fast connections without the need to pair devices. There are two types of NFC devices – passive and active devices.

1. *Passive devices:* These devices don't use an electric power source. Examples include tags attached to items in inventory and assets and credit cards.
2. *Active devices:* These devices use an electric power source. Examples include mobile phones and NFC readers used for contactless payment, physical access control, and stock taking.

Blockchain

Blockchain allows companies to execute transactions without the need for trusted intermediaries. These intermediaries are removed by recording transactions in a trusted ledger shared among the participants of a blockchain network. Blockchain, in its current form, includes a concept called smart contract that defines all conditions for a successful transaction. These smart contracts will execute without human intervention when all the requirements are met. The removal of intermediaries and automated contract execution reduces the cost and increases the speed of executing transactions.

Today, operations like insurance, traceable sourcing, and many others use blockchain. Due to the promise of efficient and low-cost transactions, blockchain transactions will appear in your industry soon, if not already. Understand it and leverage it for more efficient and lower-cost operations.

CHAPTER 7

Digital Infrastructure

The *Why* and *What* of your digital transformation must guide your choice of infrastructure. Else you will be susceptible to the hype from infrastructure companies. Understanding the digital infrastructure will prepare you to lead a purposeful digital transformation. The essential digital infrastructure components are:

- Cloud
- Internet of Things (IoT)
- Edge Computing
- Emerging infrastructure such as 3D Printing, autonomous vehicles, 5G cellular networks, and supply chain as a service

Cloud

Cloud means a lot of things to a lot of people. For the purpose of this book, I will narrow the definition of Cloud to IaaS, which allows you to rent computing and storage capacity on-demand from a service provider. You rent a whole device or part of a device and run your business apps on the rented devices. The service provider will own and operate the devices you rent.

Cloud use cases

- *Saving Capex to get extra capacity for demand surges* – Buying just the infrastructure required for average capacity and using IaaS for peak capacity is a good approach to saving Capex. For example, you can use IaaS to handle seasonal e-commerce demand.
- *Saving Capex for modernizing infrastructure* – Using IaaS when on-prem hardware becomes obsolete is one way to modernize computing, storage, and networking infrastructure without increasing Capex spend.
- *Saving Capex when expanding into new geographies* – Storing data within the new geography is sometimes critical due to data privacy rules. Due to their global presence, IaaS providers can help you to rent storage and computing capacity in the new geography.
- *Adding more IT infrastructure capacity without adding more IT staff* – The IaaS provider takes care of operations, monitoring, and maintenance of the infrastructure, allowing you to deploy more capacity without the corresponding increase in people.

Cloud myths

Myth #1: *Cloud will make you a digital leader.* Using IaaS is a choice you make whenever you have a suitable use case. Using IaaS has nothing to do with industry leadership. Being the first company in an industry to "move to the cloud" may make you a pioneer in your industry. However, leadership still depends on how your customers value your company.

Myth #2: *Regulated industries cannot use the Cloud.* There may be many restrictions on regulated industries about data that can be outside the company. However, there are many non-regulated activities where regulated industries can still use IaaS. For example, you can develop and test your applications using IaaS and deploy the final apps on-prem. Many IaaS providers offer secure computing services certified for use by specific industries. These services can be good options for regulated industries.

Myth #3: *Cloud access is always on-demand.* Renting on demand is possible, but you are not always guaranteed to get the capacity you want. For assured capacity and low cost, most IaaS users sign reserved capacity contracts with the option for burst capacity.

Myth #4: *Cloud is cheap.* IaaS can be more affordable compared to the total cost of ownership of on-prem equipment. But if you do not manage the cloud capacity diligently, you will end up paying for the capacity you are not using. For example, computing devices at the IaaS provider's datacenter will accumulate billable hours if you do not explicitly release resources from use. If you release the device, the provider can give it to another customer. If not, it is the same as you using it, although the machine was idling. So, manage your "rented" capacity diligently. Another area where you can get billing surprises is due to the amount of data that you move in and out of the provider's datacenter. Your contract will say what the data rates are, and you must pay attention to the data movement costs.

Myth #5: *You must "modernize" your apps to move to the Cloud.* It is possible to run existing on-prem apps in the IaaS provider's datacenter without any changes. This approach of moving apps to the Cloud is called "lift and shift." You will have to take care of software licensing if you move third party apps to the Cloud. Most software vendors will allow you to use your on-prem license off-prem. Sometimes the vendor may have to take specific steps or even provide you with an off-prem license in the place of your existing on-prem license.

In the case where a business process change requires you to modify a business app and move it to the Cloud, you have two options. You can update the on-prem app and do a "lift and shift," or you can "modernize" the app for the cloud environment. IaaS and on-prem app deployments differ in scale, execution environments, and data movement costs. Therefore, apps purpose-built to run in an IaaS environment (aka cloud-native apps) run and behave better in the Cloud. "Modernize" is another way of saying, "I am going to rewrite the app as a cloud-native app." The choice to modernize depends on how long the rewrite will take and if you can wait that long.

Myth #6: *Cloud-Native Apps run anywhere you want.*

Most IaaS providers bundle app development tools and building block technologies along with IaaS. The common name for this bundle is platform-as-a-service (PaaS). Developers can leverage PaaS to speed up cloud-native app development, but there is a high chance that you will be stuck with the PaaS provider for running the apps. If you are planning to use the modernized app in the IaaS environment and on-prem, ensure that going "cloud-native" will not prevent on-prem use. Most Cloud vendors are working towards the ideal state of write once and run on any cloud. The ability to write code capable of running on any Cloud will accelerate Cloud adoption.

Internet Of Things (IoT)

Tens of billions of devices will connect to the internet in the next few years. The term IoT describes the use of devices (things with sensors and actuators) sending data to a data-store over a network, followed by a process taking action based on the device data.

A good example is your video doorbell. When the bell detects motion, it turns on the video camera, notifies you, and starts streaming video to a cloud datacenter. You can remotely interact with the bell using your phone to watch the video and talk to the person in front of the bell.

Your video doorbell used your home internet gateway (WiFi Router) to alert you and stream video. Similarly, companies can also use gateway devices to collect data from many devices and send it upstream. The data from the gateway can be sent to an on-prem data-store over the company's internal network or to an IaaS data-store using a telco network.

IoT components

1. *IoT device that collects information.* The data collected can be anything that the device can sense – conditions within itself or its surroundings. Commonly collected data include – device location, low battery, component failure, motion detected, overheating, and inventory levels. The data that the device collects is up to your imagination. Collect whatever data you think will create value to the customer or your business.
2. *Network to transmit data from the IoT Device.* IoT devices do not need to use the internet; they can use any network, including your corporate network. It is also possible for IoT devices to connect directly to the IaaS data-store using 4G or 5G wireless networks.
3. *Data-store.* A scalable data-store is essential for IoT implementation.
4. *A process to take action based on the IoT device data.* An automated process to analyze the data from the IoT device is the best way to generate insights that can drive action. For example, if a soda vending machine reports that it has only three Diet Coke's left, an automated process can schedule someone to replenish Diet Coke. If the same machine reports that the inside

temperature is not cold enough to chill sodas, the automated process can schedule maintenance.

IoT use cases

IoT use cases vary from industry to industry, so use your imagination and leverage data from devices to create value to your customer and your business.

- Preventive maintenance of equipment
- Remotely turn on features for a period the customer selects and pays for online
- Usage-based billing (power meters)
- Real-time tracking of inventory and assets
- Patient monitoring
- Monitoring the power grid
- Building management (power, HVAC, lighting)
- Position items for ideal assembly yield
- Remove defective parts from manufacturing lines

An IoT enabled soda machine

Let's say you designed a soda vending machine that signals when inventory is about to be depleted. To achieve the highest possible revenue, you have to deliver the right amount of sodas to each machine as soon as the soda inventory is low. If you store the inventory depletion data from all machines, you can figure out the consumption of each soda type per day. These insights will give you the ability to stock the machine for the most revenue.

You may learn that a particular machine runs out of Diet Coke on Monday afternoons, runs out of Mountain Dew on Wednesdays, and is not selling any of the Indie brand soda. You could replace the Indie soda with Diet Coke and add more Mountain Dew when re-stocking Diet Coke on Tuesdays.

There are other types of insights you can derive from the vending machine data. If you see a sudden increase or decrease in the consumption of a particular beverage, you can investigate the cause and take the required action. For example, you get the insight that from two weeks ago, the machine seems to run out of Gatorade, which has not happened before. In your investigation, you may find that a gym has opened up across the street. You can now stock the machine with more Gatorade because this change in consumption is here to stay. Or you can talk to the gym owner and place a "gym" focused vending machine in the gym.

Hopefully, the soda vending machine example gave you an idea of the power of IoT.

IoT myths

Myth #1: *You cannot execute a digital transformation without IoT.* No rule says IoT must be part of every digital transformation. Use IoT as part of your transformation if it adds value to the customer or value to the business. A large variety of compact sensors and actuators are available today. Therefore, you can extract many types of information from devices. However, don't fall into the trap of vanity projects to make your product sound cool. Don't make up an IoT use case if the data you can collect does not provide value to you or the customer.

Myth #2: *IoT requires a big data analytics solution.* Whether you need a big data solution to go with your IoT implementation depends on the volume, velocity, and variety of the data your device sends and what you want to do with the data. Let me continue with the internet-enabled soda vending machine example to make this clear. Let's say you designed the machine to send the following:

- Heartbeat (everything is ok) – every 5 minutes
- The machine is not cold enough – on detecting condition
- Power is going down – on detecting condition

You will get a lot of heartbeat signals, but you don't need to save all the heartbeat messages. You only need to check if the heartbeat signal stops. The device sends all the other information on detecting a particular condition. The maintenance items are going to be rare events, so you will not have a high volume of data. You can discard the data after an automated process schedules maintenance.

If you have thousands of soda machines, you will get a high volume of data, and you may want to invest in big data technologies. Otherwise, it will not make sense to put a big data analytics solution in place.

Myth #3: *IoT devices are well behaved.* Many standards organizations are working towards IoT device security and interoperability. However, the current state is chaotic, and capabilities are not standard across IoT devices

One of the common bad behaviors of an IoT device is how it participates in the network. Some IoT devices generate a lot of traffic in the network, collect data flowing through the network, and leave network ports open. Such bad behavior can lead to network outages and allow cybercriminals to hack the IoT device and steal sensitive data or cause a network outage.

IoT devices have varying computing, storage, and security capabilities. This diversity makes writing IoT device software and IoT apps challenging. Also, software packages used to derive insights from IoT devices may not work with all devices. In this case, you will have to add the specific device support to the software. Major Cloud providers supply resilient and secure software for use on IoT devices along with matching IoT services on their Clouds. These stacks and services are a good starting point for most IoT implementation.

Make security a critical consideration when you select IoT devices, networks, data-stores, and analytics software. The more private and regulated the data, the more secure the IoT implementation must be.

Edge Computing

There is a lot of confusion around edge computing since there is no commonly accepted definition for it. Additional confusion ensues since edge computing is pulled into the hype of digital transformation marketing.

In the simplest definition, edge computing processes data close to where it originates or gets used.

The commonly touted uses of edge computing leverage IoT devices, but edge computing is useful without IoT devices as well. One of the first use of computing and storage at the edge was content distribution networks (CDNs). CDNs were initially used to distribute website content using computers located across the world. Due to the reduced access latency achieved by distributing websites at the edge, users got a better web experience.

Today video streaming companies like Netflix and Amazon use a CDN – their own or as a service from a CDN operator. CDNs allow video streamers to save money by not having to send many copies of the same video in different formats over the network. And viewers get a better viewing experience due to lower latency.

As in the case of IoT, do not makeup use cases for edge computing. Four factors should decide if you should use edge computing to process the data close to the source or consumer of the data.

1. *Time value of data.* If the data is not analyzed or accessed fast, it becomes useless
2. *Communication cost.* If the volume of data is so large that it is expensive to send it to the Cloud

3. *Data privacy and security.* If you must store protected private information within the required geography
4. *Service responsiveness and reliability.* If distributing functionality among many redundant edge devices will result in better service responsiveness and reliability

Edge Computing Use Cases

- Real-time safety monitoring of water, air, working conditions, power grid, and oil/gas pipelines (IoT use case – communication cost and time value)
- Real-time monitoring of equipment whose downtime is expensive (IoT use case – time value)
- Building energy management (IoT use case – communication cost)
- Processing data from oil exploration rigs before moving the Platform to a new location (non-IoT use case – time value)
- Mobile service providers (Telco) moving services to the edge to offer lower latency to customers (non-IoT – time value, service responsiveness)
- Customer authentication implemented at the edge to prevent large areas from not being able to access mobile services (non-IoT – service reliability)

- CDNs for distributing websites, video, and other streaming services (non-IoT – time value, service responsiveness, and reliability, communication cost)
- Point of sale authorization (non-IoT use case – time value, service reliability, data privacy, and security)

Edge Computing Myths

Myth #1: *Edge computing is just more responsive IoT.* In an IoT application using edge computing, a computer sitting between the IoT device and the central data-store processes the device data and generates insights. So, edge computing, when combined with IoT devices, can make some IoT apps more responsive. However, edge computing should be used in an IoT app only if one or more of the edge computing factors come into play. Besides, there are edge computing use cases that do not use IoT devices.

Myth #2: *Edge computing will replace IaaS.* Edge computing is complementary to IaaS. Edge computing allows the implementation of low latency services closer to the user and data. Only about 10% of the services need to be close to the user and data due to latency, bandwidth cost, or data privacy. IaaS is appropriate for the rest of the services.

Myth #3: *Edge computing needs 5G networks.* Edge computing is network agnostic. Communication with edge computing devices can happen over any wired or wireless network. This myth exists because 5G networks use edge computing to implement a lot of services closer to the user. 5G networks also promise higher bandwidth and lower latency communication, so eventually, 5G will be an option for connecting edge computing devices to end devices.

Myth #4: *Edge computing is only for mobile or cloud service provider use.* Mobile service providers and cloud service providers use edge computing to deploy services. But edge computing can be used by anyone who needs to process data based on the four edge computing selection factors. Since service providers offer servers or software that extend the Cloud to on-prem infrastructure, edge computing gets mistakenly associated with Cloud service providers. For example, Microsoft offers Azure Stack software, and Amazon provides Outpost servers. Such software and servers offer a Cloud-like environment on-prem and are not edge computing devices.

Emerging Infrastructure

Many infrastructure capabilities owned and vertically integrated by companies in the past are now available as a service. Manufacturing as a service, logistics as a service, customer support call center as a service, etc. have been available in the market for decades. I am not going to dive into the existing types of infrastructure. But if you are still vertically integrated, use some of these services when you plan your digital transformation.

3D Printing

As 3D printing becomes cheaper, it will be a foundational capability in manufacturing, and some manufacturing plants will switch to 3D printing plants. 3D Printing will also accelerate on-demand product personalization.

Autonomous Vehicles

Self-driving vehicles will revolutionize logistics and last-mile delivery. Fleets of autonomous vehicles and drones will become the backbone of goods transportation and last-mile distribution. Imagine an autonomous moving locker showing up in front of your home, and you opening the appropriate locker using your mobile phone. This is just like you going to an Amazon locker and opening it with your phone. The only difference is the convenience because the locker comes to you.

Supply Chain as a Service

Supply chain as a service is a very nascent digital infrastructure that offers all capabilities needed to put a supply chain in place. In this case, procurement, manufacturing, warehousing, and logistics are available on a digital platform. You will be able to choose from vendors specializing in each of these areas. The platform provider gives you a web-based interface to operate the virtual supply chain to deliver products to customers. You get to use a scalable supply chain with proven software and curated vendors. You also don't need to recruit specialists and can operate the supply chain with minimal in-house staff. The platform provider makes money by taking a cut of the monetary value of the transactions.

We are still in the early days of supply chain as a service, but it will become prevalent in the future. In the near term, leverage companies that can provide most of the supply chain capabilities under one roof. This will reduce supply chain complexity and prepare you to use supply chain as a service in the future.

5G Cellular Networks

5G is the fifth-generation cellular data network offering higher speed data with lower latency compared to a 4G network. The increased speed and network capacity, along with lower latency, aim to give a better experience to mobile users, including better streaming. The increased speed and lower latency provided by 5G networks will also allow more responsive IoT applications. And the increased capacity of the 5G networks will enable more IoT devices to connect to the network.

CHAPTER 8

Digital Operations – Art Of The Possible

Digital Operations, operating your business processes using digital data, will help you meet digital-age customer expectations and make it easy for customers to do business with you.

Digitization

The task of making the data used by your business processes available in a format that a computer can handle without user intervention is called digitization.

A long, long time ago, people used to hand out business cards, and people who got the cards would file them in business card files. When it was time to find someone's contact information, people would physically flip through the pile of cards and find the appropriate card. Some of us tried to get our collection of business cards on a computer so that we can find contacts quicker. If you ever tried to do this, you know what we had to do. Place a card in a scanner, scan the card, and run it through optical character recognition (OCR). Now the fun started. It was the early days for OCR technology, and cards came in various sizes, fonts, and formats. Therefore, OCR output had errors that required clean up. Now it was time for the next card. Fun! That was digitization.

If your data account information, invoices, etc., are stuck in paper, you will have to do a lot of work to get data digitized. However, this effort is worth it because of the benefits provided by digital operations. Digital operations:

1. *Make it easy for customers to do business with you.* To a customer, ease of doing business is second only to how well the product fits their needs.
2. *Make your business processes agile and efficient.* The productivity, agility, and effectiveness improvements make you more responsive to customers and reduce operating costs.

Cautionary Lessons

Before we explore the art of the possible using digital operations, let me caution you on a couple of topics. These cautionary words are not to discourage you from making changes; instead, they will allow you to lead a deliberate, purposeful, and successful digital transformation.

1. *Meeting customer expectations will not be enough.* A lot of digital-age customer expectations are table stakes. Simply meeting these expectations will not give you the profit growth you need. Without offering a significantly better product fit or some value that customers cannot get elsewhere, you will not get a lot of business
2. *Cost benefits will take time.* Digitizing operations always involves an initial outlay of money and will increase costs. In the long run, you may reduce costs, but until then, you will incur a cost increase. Even if you achieve the planned cost savings, it will take time to recover the money you spent.

High-level Business Processes

Your company uses a large number of processes to deliver business results, and these processes are tuned for the way you do things. The high-level processes considered in this chapter for digital transformation are the ones that have a significant impact on customer value and your costs. The high-level processes are:

- Product Development
- Supply Chain

- Sales
- Product Marketing
- Customer Support
- Finance

Information Technology (IT) – IT facilitates the operation of all the above processes, and your CIO and her representatives will be critical participants in your digital transformation. However, for the purpose of this book, I will not dive deep into IT processes.

Product Development

Critical understanding of market trends, customer requirements, and competitive positioning, along with the actual engineering of the products, belongs to this group. The following table shows how digital operations can help product development teams address customer expectations.

Customer Expectations	How to meet	Technology (or Policy)
▪ Product Fit	▪ Understand and anticipate needs ▪ Product Quality ▪ Implement product that connects back to your company (call home) ▪ Build inference engines into the product	▪ BDA (Market, customer, competition data; Call home analytics) ▪ DevOps tech ▪ Test Automation ▪ AI tasks
▪ Easy access to product details	▪ Create easy to use product documents	▪ Web tech ▪ Search

Customer Expectations	How to meet	Technology (or Policy)
▪ Mobile	▪ Mobile App ▪ Connected product	▪ App tech ▪ NFC; WiFi; 5G

▪ Usage model
  ▪ Own
  ▪ As a Service
  ▪ On-premise
  ▪ Off-premise
	▪ Licensing model
	▪ Pricing model
	▪ Usage measurement
	▪ Remote management
	▪ On-prem management
		▪ Management APIs

▪ Don't want to buy a product bundle.	▪ Change product and pricing policy	▪ Policy

▪ Pricing options
  ▪ Per-use
  ▪ Per outcome
	▪ Design to allow new pricing options
	▪ Change pricing policy
		▪ Policy

▪ Payment options
  ▪ Monthly/Yearly subscription
  ▪ Per-use
	▪ Change product to accommodate new payment options
	▪ Change pricing policy
		▪ Policy

▪ Easy replacement of the product	▪ Design principle	▪ Policy

▪ Problem handling
  ▪ Self-service
  ▪ Chat/Email
  ▪ Phone support
  ▪ Remote fixes
	▪ Design principle
	▪ Build remote troubleshooting capabilities
		▪ Knowledgebase
		▪ Policy

▪ Ability to repair, service, and upgrade with minimal disruption.	▪ Build this capability into the product ▪ Automated reminders	▪ AI tasks ▪ Policy

▪ Transparency
  ▪ Materials used
  ▪ Environmental
  ▪ Traceability
	▪ Design principles
	▪ Build traceability into product and support APIs.
		▪ Policy
		▪ NFC
		▪ RF Tags
		▪ D2D API

Supply Chain

This group of processes includes the procurement of raw materials, manufacturing, and delivery to customers. If your product is solely software, these processes handle the delivery of licenses and license management. The following table shows how digital operations can help supply chain teams address customer expectations.

Customer Expectations	How to meet	Technology (or Policy)
▪ Personalized	▪ Flexible manufacturing to personalize orders ▪ New customization partnerships ▪ Use APIs to support personalization	▪ B2B API
▪ Automated order checking for accuracy	▪ Implement this support as part of order processing	▪ e-commerce ▪ RPA ▪ AI tasks
▪ Recommendations for items needed to use the product that is missing from the order	▪ Implement this support as part of order processing	▪ e-commerce ▪ AI tasks ▪ Web tech

▪ Delivery options (with tracking)
  ▪ Instant download
  ▪ Instant online access
  ▪ Same day
  ▪ Two days
  ▪ Scheduled delivery
	▪ Implement this support as part of order processing
	▪ Establish logistics partnerships for low latency delivery
	▪ Use APIs to support automated tracking
		▪ Web tech
		▪ B2B API

▪ Mobile	▪ Mobile App	▪ App tech
▪ Ability to search for the status of in-progress orders and service requests	▪ Implement the ability to search status ▪ Push status to customers	▪ Search ▪ Web tech ▪ Mobile app tech
▪ Hassle-free product return	▪ Partnership for Reverse Logistics ▪ Use APIs to automate product return with a partner	▪ B2B API ▪ Policy

▪ Usage model
  ▪ Own
  ▪ As a Service
  ▪ On-premise
  ▪ Off-premise
	▪ Establish reverse logistics for returning equipment
	▪ Use APIs to automate on-prem aaS equipment return with a partner
	▪ Establish partnership to refurbish returned equipment to factory standard
	▪ Use APIs to automate restocking, drop-ship, and track refurbished inventory at refurb partner
		▪ B2B API

Customer Expectations	How to meet	Technology (or Policy)
▪ Recycling support	▪ Establish partnerships for Reverse Logistics ▪ Use APIs to provide automated recycling support	▪ B2B API
▪ Transparency ▪ Materials used ▪ Environmental ▪ Labor ▪ Traceability	▪ Compliance check ▪ Use the capabilities built into the product to trace the product	▪ NFC ▪ RF tags ▪ IoT

Sales

This process group handles customer engagement, customer account management, channel management, opportunity pipeline management, e-commerce, order placement (if not using e-commerce), and all activities that facilitate customers to buy your product. The following table shows how digital operations can help sales teams address customer expectations.

Customer Expectations	How to meet	Technology (or Policy)
Usage model ▪ Own/As a Service ▪ On/Off-premise	▪ Implement aaS sales process and tools	▪ Web tech
▪ Personalized	▪ Account management ▪ One-time identification – no need to repeat information	▪ CRM ▪ Data Management ▪ Single sign-on for customers
▪ Payment options ▪ Monthly/ Yearly subscription ▪ Per-use	▪ Implement payment processing application (can be a partnership)	▪ Fintech and payment processing B2B APIs ▪ Secure transactions
▪ Financing options ▪ Lease (monthly/yearly) ▪ Installment plan ▪ Instant installment plan at point of sale	▪ Implement low touch contracting ▪ Implement a loan facility with a fintech partner	▪ e-commerce ▪ Web tech ▪ On the fly contracts ▪ RPA for approvals

Customer Expectations	How to meet	Technology (or Policy)
▪ One-click ordering	▪ Implement the capability	▪ e-commerce ▪ Web tech
▪ Mobile	▪ Mobile app for ordering (single app across all processes)	▪ Mobile app tech ▪ Mobile e-commerce

▪ Payment types
  ▪ Contactless cards
  ▪ Direct transfer
▪ PayPal
	▪ Deploy appropriate technologies	▪ NFC
		▪ B2B API

▪ Don't want to buy a product bundle	▪ Exception processing ▪ Implement unbundling if needed	▪ Policy
▪ Facilitation of used product resale	▪ Partnership	▪ Web tech ▪ B2B API

Product Marketing

Best-in-class companies already use digital marketing as part of their product marketing mix. Activities related to driving awareness and preference for your products, sales enablement, demand generation, and lead management fall into this group.

The contributed pipeline is a key metric for all marketing activities. You have to answer the questions – Did the money we spent on marketing get us revenue? Did the money we spend on a particular marketing vehicle get us enough revenue? Would we do it again?

Digital marketing allows campaigns targeting customer segments. This level of individualized campaigns is very measurable and can provide attribution of a particular lead to a campaign. Airlines, retail businesses, and financial services firms are at the forefront of using digital marketing. You can learn from them. The following table shows how digital operations can help product marketing teams address customer expectations.

Customer Expectations	How to meet	Technology (or Policy)
▪ Value is well articulated	▪ Articulate value proposition	▪ Digital content ▪ Web tech ▪ Digital Ads
▪ Ability to compare with competing products	▪ Comparison engine	▪ AI tasks
▪ Personalized	▪ Content tuning ▪ Account management ▪ Recommendations	▪ AI tasks ▪ Web tech ▪ Email Service
▪ Customer testimonials	▪ Web content	▪ Digital content ▪ Web tech ▪ Digital Ads
▪ Peer reviews	▪ Web content	▪ Digital content ▪ Web tech ▪ Digital Ads
▪ Mobile	▪ Mobile App ▪ Mobile content	▪ App tech ▪ Mobile Web tech
▪ Search	▪ Promote product site ▪ Get listed in recommended products; ▪ Show up in complementary product searches	▪ SEO for visibility ▪ Paid search
▪ Easy access to product details	▪ Web content	▪ Digital content ▪ Web tech
▪ Help in establishing selection criteria	▪ Web content	▪ Digital content ▪ Web tech ▪ Digital Ads
▪ DIY videos on how to assemble, install and use the product	▪ Release short How-To videos.	▪ Digital content ▪ Web tech
▪ Transparency ▪ Materials used ▪ Environmental ▪ Labor ▪ Traceability	▪ Create awareness	▪ Digital content ▪ Web tech ▪ Digital Ads

Customer Support

This group of processes helps customers use products effectively and offers services to make it easy to address the problems encountered while using your product.

Consumers are now used to hassle-free customer service. The fundamental rule for customer support is that your team must not waste the customers' time. Track customers' personal information, transactions, issues, and preferences to make interactions efficient and save customers' time. Customers get annoyed when they have to provide account information many times when they get transferred between different customer support teams. Once customers provide their data to a company, they expect that everyone they come into contact with at the company will have that data, so make customer profiles available to all the people who will interact with customers. Digital-age customers prefer to resolve issues using web interfaces, and they get annoyed if it takes many attempts to solve a problem.

The following table shows how support teams can use digital operations to meet customer expectations.

Customer Expectations	How to meet	Technology (or Policy)
▪ Problem handling ▪ Self-service ▪ Knowledgebase ▪ Chat/Email ▪ Phone support ▪ Remote fixes	▪ Multichannel customer contacts ▪ Staff trained in remote troubleshooting ▪ Ability to book and change service online ▪ Track technician arrival time ▪ One-time identification – no need to repeat information	▪ Single sign-on for customers ▪ Knowledgebase ▪ Web tech ▪ Digital Content ▪ Asset/Personnel tracking system ▪ IoT
▪ Hassle-free product return.	▪ Automated approval ▪ Supervised approval	▪ Policy ▪ RPA
▪ Ability to repair, service, and upgrade with minimal disruption	▪ Provide guidance	▪ Web tech ▪ Digital Content
▪ Self Service	▪ Provide content for troubleshooting and fixing issues	▪ Web tech ▪ Digital Content ▪ Knowledgebase ▪ Search
▪ Mobile	▪ Mobile App	▪ App tech
▪ Assembly and Installation	▪ Partnerships	▪ B2B API
▪ Social	▪ User Community	▪ Web tech ▪ Knowledgebase

Finance

The activities of interest here are the policies that govern financial transactions with customers and partnerships you establish with financial institutions and fintech companies. The finance team must establish payment terms, product financing, and partnerships. If not, you will be surprised by the impact of these on your profit.

The following table shows how finance teams can use digital operations to meet customer expectations.

Customer Expectations	How to meet	Technology (or Policy)
▪ Usage model ▪ Own ▪ As a Service ▪ On-premise ▪ Off-premise	▪ aaS Revenue model and financing for on-prem aaS ▪ Credit checks for on-prem aaS	▪ Policy
▪ Payment types ▪ Contactless card ▪ Direct transfer ▪ PayPal	▪ Approve these payment type ▪ Negotiate transaction fees	▪ Policy
▪ Financing options ▪ Lease (monthly/yearly) ▪ Installment plan ▪ Instant installment plan at point of sale	▪ Approve financing options ▪ Establish financing Partnership	▪ Policy
▪ Pricing options ▪ Per-use/ transaction ▪ Per outcome	▪ Pricing policy approval	▪ Policy
▪ Payment options ▪ Monthly/ Yearly subscription ▪ Per-use	▪ Pricing policy ▪ Revenue model ▪ Transition timing ▪ Investor education	▪ Policy
▪ Hassle-free product return	▪ Establish a service accounting model to handle no questions asked returns	▪ Policy

Cost Savings With Digital Operations

Although getting your business processes to digital readiness will take time and money, digital operations will provide cost savings due to increased productivity and efficiencies. As part of the pre-work for preparing the roadmap for your digital transformation, you must challenge each of your teams to create a potential cost savings table similar to the tables in this chapter. What are the digital technologies of interest in their functional area? How can they exploit the technologies of interest to reduce costs?

In addition to productivity and efficiency gains, you should also consider adopting cost-saving services like predictive sourcing, spend analytics, etc. New and innovative technologies and services come up every day to help reduce costs in all processes. Keep an eye out for cost-saving innovations.

Digital Operations And Cybersecurity

With digital operations, you collect data at all touchpoints with customers, suppliers, and employees. Besides, you will generate a large amount of data about your products and business processes. You must protect this valuable data from cybercriminals. Therefore, cybersecurity is an overarching requirement in the digital age.

In addition to protecting your data, pay special attention when you select payment services and fintech partnerships. If these partners get hacked, you will lose your reputation and will face significant financial loss.

PART 3

The When, Who, and How

CHAPTER 9

Map Your Digital Journey

The sequence in which you execute each of the business process changes decide the *When* of your transformation. Creating the roadmap for your digital transformation is a significant resource allocation exercise across many business processes, under many constraints. The different types of changes, the different types of constraints, and impact-based prioritization make the selection of change initiatives complicated.

Types of Changes

Reviewing the changes required to ensure a successful digital transformation is a good start for the resource allocation exercise we are about to start.

1. *Changes to improve customer experience.* The *Why* of your digital transformation is the changes you have to make to improve value to your customers.
2. *Changes to reduce costs.* When you start making customer experience changes, you may have to reduce costs to fund the essential customer value changes and stay profitable.

3. *Changes to mitigate technology or infrastructure obsolescence.* You will come across many changes forced upon you due to technology or infrastructure obsolescence

Constraints you work under

In addition to picking the changes and scheduling them, you have to identify the constraints under which you operate.

1. *Available funding.* The first order constraint in most companies is the availability of funding. If you cannot fund the transformation projects, they will fail.
2. *Available talent.* Finding the requisite talent to execute the transformation is another major constraint for most companies.
3. *Execution time.* The time to execute projects is also a constraint to consider when embarking on your digital transformation.

The Impact Filter

When you apply the impact filter, your goal is to select change initiatives based on the impact of the change on the business. The impact of a change is measured as the benefit you will get if you implement the particular change. The impact can be an increase in revenue, profits, or relevance, or a reduction in loss or risk. The following are some examples of impact statements for the three types of changes:

- *Customer value improvement change*
    - Prevent $50M revenue loss
    - Increase revenue by $100M
- *Cost-saving type change*
    - Save $2M per quarter
    - Reduce product cost by 10%
- *Obsolescence change*
    - Less vulnerable to hackers
    - Reduce service downtime

Give higher priority to changes with the highest impact on customer value, cost-savings, and obsolescence in that order. When you finish this exercise, you will have three lists prioritized by business needs. However, you have to apply more filters to select initiatives from among the three lists.

The Execution Time Filter

The effort required, the unfamiliarity of the tasks involved, and external dependencies cause some projects to take a long time to complete. My advice is not to include projects that will take longer than 12 months into the initial cohort of initiatives. This advice gets challenged with – "What if the change with the most significant impact takes more than 12 months to implement?"

The reason for this advice is because you need significant momentum and short-term wins to keep your team charged up. If most of your change initiatives will not complete for a year or more, you will not be able to sustain the required momentum. You can, of course, make exceptions and add a few longer-term projects into your initial cohort of changes.

Another approach for adding long-duration but critical projects into the digital transformation roadmap is to introduce the concept of Time Horizons.

- *Horizon 1:* Initiatives that can be completed in less than 12 months and can start immediately
- *Horizon 2:* Initiatives that can be completed in less than 12 months but can only start later due to talent or funding availability
- *Horizon 3:* Initiatives that take more than 12 months to complete

Introducing time horizons to your roadmap will allow you to create momentum using Horizon 1 and 2 initiatives and, at the same time, showcase the important long-duration projects using Horizon 3.

The Talent Filter

One of the critical items that will make or break your digital transformation is how you address the staffing of your change initiatives. I have heard many complaints about inadequate funding for projects. Lack of funding forces teams to be creative in partnering, improving productivity, and eliminating cruft to reduce the funding needed. However, something that happens a lot and does not get much attention is that execution teams sometimes ignore execution capacity.

The following common-sense items are often overlooked in talent allocation.

- Every project needs a minimum headcount of specific skills to start the project and can ramp to the full headcount over time
- New hire funding must be provided such that the hiring can happen before the project starts

- Hiring gets delayed due to the high demand for a specific skill
- The project schedule assumes that a new hire is productive on day one

You can use a table like the one below to assess the start of each initiative based on talent availability.

Change	Skill	Count Available	Count Needed	Count Needed to start	Possible start date
Change #1	Skill 1*	10	12	5	Now
	Skill 2	1	5	0	Now
	Skill 3	3	7	3	Now
Change #2	Skill 4*	0	10	5	+6 month
	Skill 5	2	8	4	+3 month
	Skill 6	5	7	4	Now
Change #3	Skill 6*	5	4	3	Now
	Skill 7	2	8	2	Now
	Skill 8	4	5	3	Now

**Specialist skill needed to start the project*

The example uses a small sample of skills, but the more change initiatives you have, the larger this table gets. Sometimes, many initiatives will need a particular skill. Pay attention to how your account for skills in this table. In the example above, the available count shown in each row is the total count available for each skill and not the count available for each initiative.

In this example, Change #1 is ready to start since Skill #2 is not a start limiting skill, and you have the minimum count for Skill #1 and Skill #3. You cannot start Change #2 for six months because the minimum count of Skill #4 can be hired and made available only six months from now. Since Change #2 cannot start, Skill #6 is now available for Change #3, and you can start that initiative.

Just because a project cannot start because of a lack of skills does not mean that you ignore the project. You have to start the hiring process because you must hire, onboard, and train people ahead of their contribution to an initiative. As you hire people for initiatives with a later start date, the new hires will add expenses to the project while they get ready to contribute to the initiative.

Remove table rows that belong to the first cohort of initiatives that can be started immediately and update the rows of the remaining change initiatives with the headcount available for each skill.

Change	Skill	Count Available	Count Needed	Count Needed to start	Possible start date
Change #2	Skill 4*	0	10	5	+6 month
	Skill 5	2	8	4	+3 month
	Skill 6	1	7	2	+3 month

**Specialist skill needed to start the project.*

The new table will tell you what talent count it will take to start the next cohort of changes and when.

The Funding Filter

Next to talent, funding is another major constraint. You will have to drive all the change initiatives within your expense envelope. You may be able to re-allocate extra funds and headcount to the change initiatives from the overall envelope through cost savings and productivity improvements. However, cost savings and productivity improvements take time, and in the meanwhile, change initiatives will starve.

You must fund based on your convictions and distribute funds disproportionately, favoring the way you want to impact the industry. I prefer the following allocation because it fits my priorities.

1. 50% for customer-value initiatives (security – fund as needed but try to keep below 5%; offense 25%; defense 20%)
2. 30% for cost savings initiatives
3. 20% or less for obsolescence initiatives

The 20% or lower funding for obsolescence initiatives does not mean there is very little funding for technology and infrastructure changes. In addition to the dedicated obsolescence funding, customer-value and cost-saving initiatives will also provide money and people for the required technology and infrastructure changes. Allocating funds this way ensures that the required business process changes will lead to changes in business apps, technology, and infrastructure, instead of being led by technology hype.

Creating The Change Roadmap

Step 1: *Create three separate change lists, one for each type of change.* Each change initiative must have the following information:

- Impact (revenue and profit)
- Current State (business processes and outputs)
- Future State (business processes and outputs)
- Details of what will change (policies, apps, technologies, and infrastructure)
- The cost impact of the change
- Cash management impact of new policies
- The time required to cut over to new policies
- Non-headcount funding required per quarter
- Execution start date

- Execution time in months
- Headcount required per skillset (see the format in the Talent Filter section)

Step 2: *For each list, apply the impact filter to get your priority within a list.* The following picture shows the application of the impact filter to prioritize the list.

	IMPACT		
	High	Medium	Low
Customer value initiatives ➡	List of initiatives	List of initiatives	List of initiatives
Cost saving initiatives ➡	List of initiatives	List of initiatives	List of initiatives
Obsolescence initiatives ➡	List of initiatives	List of initiatives	List of initiatives

Step 3: *Prioritize among the three types of changes.*

Priority 1: Value to the customer – Cybersecurity, offense, and defense in that order

Priority 2: Cost savings

Priority 3: Technology/Infrastructure obsolescence

The following picture shows how you can apply this prioritization to the list from Step 2.

	IMPACT		
	High	Medium	Low
Customer value initiatives ➡	Priority 1	List of initiatives	List of initiatives
Cost saving initiatives ➡	Priority 2	List of initiatives	List of initiatives
Obsolescence initiatives ➡	Priority 3	List of initiatives	List of initiatives

Step 4: *Apply the execution time filter on the prioritized list from Step 3 to move some projects to Horizon 3.*

Step 5: *Apply the talent filter to the list from Step 4.* In this step, assume that you can fund the talent. Your goal is to identify when projects can start based on talent availability.

Step 6: *Apply the funding filter to the list from Step 5.* The goal is to select the first cohort of initiatives from each list. After this step, your selection matrix will be as shown in the following picture.

	TALENT & FUNDING FILTER		
	Can Start Now	**Can Start in 3 months**	**Can start in 6 months**
Priority 1 ➡	List of initiatives	List of initiatives	List of initiatives
Priority 2 ➡	List of initiatives	List of initiatives	List of initiatives
Priority 3 ➡	List of initiatives	List of initiatives	List of initiatives

Step 7: *Apply the impact filter one more time to the projects remaining in the list.* This step allows you to check for impact across the three lists and ensure that lower impact items do not get funded just because it had a higher priority on a lower value list. For example, you may deprioritize an obsolescence initiative to support more cost savings initiatives.

Attempting to do a cross-list impact analysis earlier than Step 7 will be frustrating for you and your team. There will be too many projects, and you will be spending time on a lot of projects that will get filtered out by other filters.

Step 8: *Place each change initiative that remains in the* list after Step 7 into its appropriate Horizon. At the end of Step 8, your initiative scheduling will be as shown in the following picture.

	PRIORITY		
	Priority 1	Priority 2	Priority 3
Horizon 1 ➡	▪ All immediately start-able initiatives	▪ 50% of immediately start-able initiatives	
Horizon 2 ➡	▪ Initiatives start-able in 3 months	▪ 50% of immediately start-able initiatives	▪ 50% of immediately start-able initiatives
Horizon 3 ➡			▪ 50% of immediately start-able initiatives ▪ Long execution time initiatives

Step 9: *Create project plans for each initiative with monthly progress milestones in each project plan.* Combine the plans for all initiatives into an overall transformation roadmap.

Resolve To Fund Your Initiatives

Creating the change roadmap, the *When* of your digital transformation, is a very complex resource allocation exercise, and you have to make these changes while operating the business. One of my mentors used to say, "It is like repairing an airplane engine while the plane is in flight." This analogy is very apt, and it stuck with me.

I used to joke that doing the in-flight repair is more doable on a four-engine aircraft with engine redundancy than on a single-engine plane. I meant no disrespect to my mentor. It is tough to fund change if you are surviving on a shoestring budget and doing only the bare minimum to be in business. In this case, change initiatives will fail since operating the current business model consumes all the money and people. However, in a larger organization with tens of millions of dollars in budget and a whole bunch of projects, you will be able to find funding and people if you have the resolve.

Pace Yourself

You may want to start some projects immediately, but prerequisites will gate many change initiatives. Similarly, change management needs can delay the actual start time of an initiative. You should schedule initiatives when they can start instead of forcing them to start right away.

Your sense of urgency and drive to get things done may go into overdrive during roadmap creation but pace the change initiatives because you need:

- To find funding and resources within your operating budget
- A small enough set of change initiatives to start with so that you and your leadership team can pay adequate attention to the initiatives
- Momentum for your change initiatives to engage and excite your team

CHAPTER 10

People Strategy In The Digital Age

Whenever you change how work gets done, especially the adoption of new technologies, you will run into talent mismatch. Your loyal troops may not be able to help you take your business to where it must go. It is sad but true that mismatched talent has no place in your company. On top of this dilemma, the adoption of automation is accelerating rapidly. Automation will cut many jobs, and your digital transformation may eliminate the need for some of your team members. There is one more piece to the people puzzle when you lead a public company. You have to get your board of directors to buy into your transformation.

Leadership Mindsets

The pace at which change is happening in the business world challenges how you organize your business. The efficient organization structure you honed to perfection is going to hold you back. The following analogy will make this clear. To speed up product development, engineers use a modular approach. Product developers pull together interoperable modules that provide specific functions to get products out fast. To handle rapid change, you will have to do the same with your organization. You may have to bring together people with the skills needed to respond to a specific threat or an extraordinary opportunity. Yet another combination of skills may be needed for the next threat or opportunity you face.

One significant challenge you will face if you try to reconfigure teams is that functional and business unit fiefdoms will fight this type of modular team building. You must create a culture where leaders are comfortable configuring resources and even themselves for the best possible outcome. If the leaders should be re-deployable, they need a broad understanding of your business, not just about one specific part. In this context, some of your leaders will not be able to serve the company well. Look for the following mindsets in your leaders:

- Entrepreneurial spirit
- Not afraid to change
- Not afraid of technology
- Excellent change management skills
- Excellent cross-organizational skills
- Curious enough to learn all aspects of the business

- Equally comfortable as an individual contributor and leading a large team
- Persistence – leaders cannot fold at the first signs of resistance; they must find a way to achieve goals

You are in good shape if you have a whole stable full of leaders with these mindsets. If not, you have to get the right leaders. Hopefully, you need to recruit only a few new leaders.

Skillset

Finding the skill sets you need is your next big challenge. You will realize this when you get to the change initiative selection stage or when your team shares potential initiatives with you. If your plans call for a lot of digital technology skills, you will have a tough time executing the plan.

Due to significant demand, it is challenging to recruit data-related skills (analytics, data science, big data engineering) and AI-related skills. The smaller number of applications you have to develop in-house, the better your chances of success. In some cases, you can rely on an IaaS provider to offer base technologies. However, even in this case, you will need people who can use those technologies to build apps and use the apps. Recruiting even this smaller count of people is still going to be a tough challenge.

If your plan calls for a sizeable team with digital technology skills, ask your HR leader to prepare a digital competency roadmap using the following table as a starting point. Once you have your competency map, you should ask your HR leader to drive a focused and well-monitored recruitment process.

Skill	Count Available	Count Needed	Required By Date	Talent Acquisition Plan
Skill 1	10	12		
Skill 2	1	5		
Skill 3	3	7		
Skill 4	0	10		
Skill 5	2	8		
Skill 6	5	7		
Skill 7	2	8		
Skill 8	4	5		

If it will take a long time to hire the people you need, think about collaborating with other companies to share resources or acquihiring (buying a company just for the talent). Partnerships and acquisition have their challenges, but those may be the only way to transform successfully.

Automation

Automation of work brings many challenges to talent management. If your transformation eliminates a lot of jobs, you have to work towards a soft landing for the people who will lose their jobs. Can you train and deploy these folks elsewhere in the future state? Can you find employment for these folks in another company that may need their skills? Ask your HR leader to prepare a list of the positions that will be eliminated and a plan to find soft landings for the people who will depart.

Team engagement in the face of automation is hard to achieve. If you are automating a process using an automation technology that has not proven itself, you must pay special attention to team engagement. The team whose job will be eliminated knows that they will be out of the company when the technology proves itself. The instinct for the group will be to jump ship when they get their chance. Finding soft landings within or outside your company will help you in situations like this.

Convincing Your Board Of Directors

If you lead a public company, you must get your board of directors to approve significant changes. Use the *Why,*

What, Where, and *When* of your digital transformation to get buy-in from your board.

In most technology companies, the board of directors has an *Innovation and Technology Committee*. This committee acts as an interface to the rest of the board members and reviews the innovation portfolio, technical risks, and product development programs. Using this model, work with your Chairman and establish a *Digital Committee* to help you get buy-in from the full board. If your board has digital-savvy directors, you can put together a digital committee quickly. If not, work with your Chairman to recruit a few digital-savvy directors when the opportunity avails itself.

Most board members have extensive business experience, so non-digital directors will not pay attention to digital directors who do not have relevant business experience. Ideally, your digital directors have general management backgrounds with customer engagements across industries and exposure to digital technologies and infrastructure. Digital directors who are savvy about digital threats and opportunities can help communicate the necessity of your initiatives to the rest of the board.

Work The Way You Live

Earlier, we covered customer expectations in the digital age. Everyone in your team has the same consumer-level expectations when they deal with their company. Many SaaS solutions with intuitive consumer-style interfaces are available to address your team's needs. These tools will allow you to deploy company-specific social networks, hassle-free expense reporting, and self-service portals. Consumer type messaging and blogging makes it easy to communicate in ways that everyone uses in their life outside of work. The information transparency possible with these team solutions will be of great use as you progress in your transformation. Add these types of changes to your transformation effort, and your team will appreciate the changes you are making for them.

CHAPTER 11

Lead The Change

It takes great focus and disciplined execution to pull off a successful digital transformation. Purposeful digital transformations cut across many departments where teams are taking on unfamiliar tasks, technologies, and infrastructures. In this chaos, they have to make tradeoffs across different groups and projects. In this environment, "transformation by committee" will not work. The person driving the change must have the clout to make things happen in this chaos. You are the one with the most clout, and this means you must lead the transformation.

Understanding why change initiatives fail is a crucial requirement to ensure a successful digital transformation. Change initiatives fail if you don't:

- Know what you need
- Ask for what you want
- Convince people that the change is good for them
- Apply enough money and people to the initiative
- Monitor execution and take corrective actions
- Deal with bad behaviors

Be Clear About What You Want

Focus on a few clear goals

It is very tempting to set a lot of goals for your team. The overarching thought, in this case, is that the more goals you have, the more people will stretch and achieve those goals. However, it is challenging for people to focus if they deal with seventeen high priority goals. So, focus on three to four goals, top-priority goals.

Over Communicate

Everyone in your team must know what they have to do. Communicate your goals and plans at every opportunity you get. You may get tired of repeating the message many times, but people need to hear this message many times before they internalize it. Repetition will ensure that your team understands why the goals are important and what they must do to achieve the goals. Articulate the changes, the need for change, and the roadmap from now to the future state.

You want your team to achieve your goals without any compromises. So, when talking about goals, get rid of the words – "should," "try," and "if possible." These words leave outcomes to the discretion of people working towards the goal. Instead, use the word MUST.

Establish a common vocabulary

Establishing a shared vocabulary for the transformation is essential. The goal of the shared vocabulary is to make your requests clear and for everyone to align with the process you want to follow. The shared vocabulary will also make sure that you and your team members mean the same thing when you use keywords. You can use the terms in this book or modify them to match your team's vocabulary.

Having a common vocabulary will also allow you to explain the reasons for pacing the initiatives into three time horizons. People should not worry about working on Horizon 2 and Horizon 3 projects. They should also not feel frustrated that their initiative is not in Horizon 1.

Assign clear accountability for results

There is nothing better than associating a person's name to a specific goal to instill accountability. Vague statements like, "We will survey ten customers by the end of the year," is not going to make team members accountable. Instead, stating the goal as – "Jill will be responsible for surveying ten customers by the end of December," will ensure that Jill knows she is accountable for the result.

All action items coming out of your meetings must have a person's name and a due date attached to it.

Convince Your People

Get buy-in

You need everyone in the organization to feel the need to change and have the same sense of urgency for the change as you. To get to that level of buy-in, you must start by getting buy-in from your senior leaders and the opinion leaders who are not necessarily your senior leaders. Getting these folks involved in the pre-work for the transformation is critical for your success. Since there is a lot of pre-work to do, delegating the work will also help move the pre-work along faster.

You can also engage your high potential talent to do the pre-work. Participating in the identification and selection of changes to push the company forward will be good for them, and they, in turn, will get good exposure to senior leaders. The next set of folks you need to convince are the first-line managers. These front-line leaders are the ones who have real control over execution, hiring, schedule, and costs. If these leaders do not have a clear understanding of your goals and don't have guiding principles to make decisions, their daily choices will derail the transformation.

Senior leaders, opinion leaders, high potentials, and first-line managers will help you get the next level of buy-in. These leaders can be additional resources to amplify your messages in the organization. Ask each leader to workshop what this transformation means to them and their teams. The words they develop will be excellent messages for them to use with their teams.

Deal with the lizard brain

As soon as people become aware of a change, their lizard brain takes over and asks many questions. The four common questions are:

1. Is this change dangerous for me?
2. What is in this for me?
3. Is all this work worth the benefits I get?
4. What happens if this initiative fails?

There are no magic answers, but you must find the appropriate, personalized answers for the key leaders whose buy-in is essential for the success of your digital transformation. They can then develop responses relevant to their team members.

Communicate and be visible

Congratulations!! You are now ready to launch your transformation to the broader team. Make sure you communicate the results you want and include messages for the lizard brain. Over-communicate and have your leaders over-communicate.

The initial push of messages will not be enough to maintain momentum for the transformation. You have to keep up the drumbeat and show progress. Progress and hitting milestones pump up people, so communicate status broadly after each monthly review. Celebrate successes and showcase the heroes. And finally – be visible. Visible leadership by you and your senior leaders is a must for your success. Make sure that your team sees your focus, determination, enthusiasm, and frustrations about the change.

Fund YOUR Transformation

A lack of resources (money and people) is the primary reason why projects fail to meet expectations. If you start a digital transformation, you must be willing to find the funding for it. As a leader, this is your responsibility. You and your team have to work within the constraints of your expense envelope. So, it is not enough for you to just set the vision; you must help your team fund the initiatives. First, you must clarify that you expect your team to execute the transformation within the existing expense envelope. Second, they will need your help reprioritizing current projects or getting seed funding for cost savings initiatives. If the cost-saving initiatives deliver as expected, you will be able to fund change initiatives in a quarter or two.

Monitor The Execution

Insist on written plans for all significant goals

Even if people are fully engaged and bought-in to your goals for the digital transformation, it is only natural that they get distracted by their day to day activities. So, all significant goals must have written plans for achieving them. The written plan must translate your team's ideas for achieving a particular goal into a milestone-based plan. Split each program into a series of milestones that are roughly a month apart. These monthly milestones will help your team members stay focused, give you leading indicators for issues that can derail the project, and take corrective actions.

Review progress against written plans monthly

At each monthly review, you must assess progress against the written plan. You have to realize you are failing before you can take corrective actions. Therefore, you have to develop leading indicators for issues that can derail your change initiatives.

The milestone review is the best place to test critical assumptions about the project, detect issues, take corrective actions, and even redirect an initiative. Here are some suggestions on using a milestone review to develop leading indicators for the success of an initiative:

- Ask the project team members what issues the team is facing, including dependencies that are chronically hampering execution.
- Monitor hiring every month. Even if you provide funds for headcount, project teams sometimes won't be able to hire as planned. Inadequate staffing is a leading indicator that the project will fail or finish late.
- Check project funding every month. In some cases, the funding you approved will not reach the project teams. Lack of adequate funding is a leading indicator that the initiative will go off the rails. You must tell the

person accountable for the initiative that you expect her to tell you if the funds are not released on time.

- Compare the initial assumptions about the market, resource availability, quality, expenses, and customer acceptance to the current assessments.

Your goal should be to build a deep understanding of what is happening with the initiative. Since you are developing this understanding while the key members of the initiative are with you, the whole team understands the initiative better.

Your questions will make the team think about the topics you brought up during the project, especially when they prepare to meet you for the next milestone review. People typically want to look good in front of their boss. Therefore, they will come well prepared for the next milestone review. It is not the preparation for the milestone review that will help achieve the goal. Instead, your questions will make the team develop a better understanding of the initiative and about the things that can go wrong, and they will start paying attention to those details.

Hold people accountable

Naturally, some things will not go as planned. The person who is accountable for delivering the results must closely monitor each initiative's progress towards its next milestone and make sure that everything possible gets done to achieve the planned intermediate goals by the milestone date.

Make sure that leaders and managers in your team know that you are serious about commitments. A milestone review is not the place for people to complain about their problems or play a victim to circumstances. If you see deviations from the plan, you must confront the people who did not meet the goals. You must, in clear terms, let the people who miss their goals know that missing commitments is not acceptable. Once you have confronted the people who did not achieve a goal, ask the team what they learned from missing the milestone and how they will prevent it from happening in the future.

Observe how your first line managers behave and coach them if needed. First-line managers are critical in keeping commitments – if something slips, they should be asking – How do we get this back on track? And not play the victim.

Deal With Bad Behaviors

Gaming the system

Team members may try to game the system by inflating the impact of their initiatives to get higher priority. All of a sudden, someone's favorite cost savings initiative has the potential to take out 50% of your operating costs, and someone else's favorite CRM system promises to double your revenue.

If you are not sure about an estimate, talk to a mentor or board member who has experience with such initiatives, and you pressure test it. You must have very direct conversations with folks who try to game the system. Sometimes the intention will not be bad; they just want to get their initiative going. If the initiative is in your selected list and you have it scheduled for a later time, your discussion may be enough to get these folks from trying to game the system.

Even if you do not detect the gaming early on, you have a great way to end such behavior during the monthly reviews. During these reviews, you will get early indicators if someone puffed up an initiative just to get it selected. Hold the person accountable and dole out some consequences.

Pocket Vetoes

Some leaders will not take the required actions before the time runs out. Delaying actions is terrible behavior, especially coming from a leader in your organization. Indecision is a debilitating weakness for people in leadership roles, and leaders who cannot decide must be removed from your team. But pocket vetoes are pure evil. The leader knows what she must do but is showing her disagreement by not taking action. You must have the tough discussions when necessary to remove the conflict or remove the person.

Negativity

Negativity is a poison that will kill your initiatives. Keep a close watch on negativity emanating from your leaders. Parochial leaders will not be open to change and will engage in negative behavior, including turf wars. These folks get defensive and negative if the transformation requires them to provide resources to other teams or take on new work. People watch leaders, so negativity from leaders spread fast. You have to either get these leaders onboard the transformation or move them out of the company.

Made in the USA
Monee, IL
06 November 2026

68002971R00094